Practical Induction Heat Treating

Richard E. Haimbaugh

The Materials Information Society

Materials Park, Ohio 44073-0002
www.asminternational.org

Delta College Library
April 2004

Copyright © 2001
by
ASM International®
All rights reserved

No part of this book may be reproduced, stored in a retrieval system, or transmitted, in any form or by any means, electronic, mechanical, photocopying, recording, or otherwise, without the written permission of the copyright owner.

First printing, December 2001

Great care is taken in the compilation and production of this book, but it should be made clear that NO WARRANTIES, EXPRESS OR IMPLIED, INCLUDING, WITHOUT LIMITATION, WARRANTIES OF MERCHANTABILITY OR FITNESS FOR A PARTICULAR PURPOSE, ARE GIVEN IN CONNECTION WITH THIS PUBLICATION. Although this information is believed to be accurate by ASM, ASM cannot guarantee that favorable results will be obtained from the use of this publication alone. This publication is intended for use by persons having technical skill, at their sole discretion and risk. Since the conditions of product or material use are outside of ASM's control, ASM assumes no liability or obligation in connection with any use of this information. No claim of any kind, whether as to products or information in this publication, and whether or not based on negligence, shall be greater in amount than the purchase price of this product or publication in respect of which damages are claimed. THE REMEDY HEREBY PROVIDED SHALL BE THE EXCLUSIVE AND SOLE REMEDY OF BUYER, AND IN NO EVENT SHALL EITHER PARTY BE LIABLE FOR SPECIAL, INDIRECT OR CONSEQUENTIAL DAMAGES WHETHER OR NOT CAUSED BY OR RESULTING FROM THE NEGLIGENCE OF SUCH PARTY. As with any material, evaluation of the material under end-use conditions prior to specification is essential. Therefore, specific testing under actual conditions is recommended.

Nothing contained in this book shall be construed as a grant of any right of manufacture, sale, use, or reproduction, in connection with any method, process, apparatus, product, composition, or system, whether or not covered by letters patent, copyright, or trademark, and nothing contained in this book shall be construed as a defense against any alleged infringement of letters patent, copyright, or trademark, or as a defense against liability for such infringement.

Comments, criticisms, and suggestions are invited, and should be forwarded to ASM International.

ASM International staff who worked on this project included Veronica Flint, Manager of Book Acquisitions; Bonnie Sanders, Manager of Production; Nancy Hrivnak, Copy Editor; Kathy Dragolich and Jill Kinson, Production Editor; and Scott Henry, Assistant Director of Reference Publications.

ISBN: 0-87170-743-8
SAN: 204-7586

Library of Congress Cataloging-in-Publication Data

Haimbaugh, Richard E.
Practical induction heat treating/Richard E. Haimbaugh.
p.cm.
Includes bibliographical references and index
1. Induction hardening. I. Title.
TN672 .H24 2001   671.3'6—dc21   2001041267

ASM International®
Materials Park, OH 44073-0002
www.asminternational.org

Printed in the United States of America

TN 672 .H24 2001

Haimbaugh, Richard E.

Practical induction heat treating

To my father, Omer, who started my interest and training in induction heat treating; my brother Dave, who does impossible things with induction coils; my brother, Kurt, who has always asked good questions; and my wife, Carol, who gives me inspiration.

ASM International Technical Books Committee (2000–2001)

Sunniva R. Collins (Chair)
Swagelok/Nupro Company
Charles A. Parker (Vice Chair)
Allied Signal Aircraft Landing Systems
Eugen Abramovici
Bombardier Aerospace (Canadair)
A.S. Brar
Seagate Technology
Ngai Mun Chow
Det Norske Veritas Pte Ltd.
Seetharama C. Deevi
Philip Morris, USA
Bradley J. Diak
Queen's University
James C. Foley
Ames Laboratory
Dov B. Goldman
Precision World Products

James F.R. Grochmal
Metallurgical Perspectives
Nguyen P. Hung
Nanyang Technological University
Serope Kalpakjian
Illinois Institute of Technology
Gordon Lippa
North Star Casteel
Jacques Masounave
Université du Québec
K. Bhanu Sankara Rao
Indira Gandhi Centre for Atomic Research
Mel M. Schwartz
Sikorsky Aircraft Corporation (Retired)
Peter F. Timmins
Risk Based Inspection, Inc.
George F. Vander Voort
Buehler Ltd.

About the Author

Dick Haimbaugh's first practical experience with induction heat treating occurred in 1946 when he was eleven years old. With the purchase of two war surplus General Electric radio frequency induction heaters, his father had started a commercial heat treating company to specialize in induction heat treating. After pestering his father for work, Dick was shown how to load hub caps into a coil, push the "on" button, and then remove the hub caps from the coil when they had been induction heat treated. He took about 2 hours to induction anneal 250 hub caps.

He worked for his father through high school and college. In his senior year in college where he was majoring in metallurgical engineering, and while working as a lab assistant for graduate students, Dick asked his father for help with the coil design for a spark gap unit. The coil pulled so much power that the cables to the power supply started smoking.

Following graduation from the University of Illinois, Dick worked for a short time for Allison Division of General Motors. Then during his Army service, he worked as a mechanical engineer in the Army Rocket Guided Missile section at Redstone Arsenal. Upon leaving the Army, Dick returned to work for his father while also attending the University of Chicago where he earned an MBA degree. Since that time, he has been involved in all aspects of commercial induction heat treating, with active participation as a lifetime member of ASM International, chairing and participating in various ASM committees, working with Handbook Committees, as well as contributing to several of the MEI courses.

Over the years, Dick has trained the personnel at Induction Heat Treating Corp. in the knowledge needed for commercial heat treating production, and he has consulted for various companies including General Electric and the Center for Metals Fabrication. He has participated as an induction heat treating expert in two Heat Treating Conferences sponsored by ASM International and plans to remain active in the industry.

Preface

Throughout the years, many books and articles have been written about induction heat treating. In the author's opinion, *Induction Heat Treatment of Steel*, Lee Semiatin and Dave Stutz, 1986, provides the best combination of induction heating and metallurgical theory to date.

There are many practical aspects that the books to date do not cover. The author's company has the experience of processing more than 20,000 orders a year in commercial induction heat treating. This book is written to complete the tie-in of the metallurgy, theory, and practice of induction heat treating from a hands-on explanation of what floor people need to know. Explanations contain language and terms that need to be understood. Operating information and a progression from process analysis to standards and quality control are presented.

The early chapters, 1 through 7, provide explanations of theory to the detail that the author feels is needed in order to understand induction and the metallurgy of induction. Chapters 8 to 10 deal with production aspects of induction. Chapter 11 reviews and presents a process for analysis of applications, including selection of frequency, power requirements, and the selection of different types of fixturing to meet production requirements. Chapter 12 discusses standards and inspection for induction, while Chapter 13 deals with identification and resolution of problems found with induction hardened parts. The final chapters discuss quality control and maintenance.

The appendixes are meant to help more with design information and include some charts and data to help with production including tempering curves and hardenability curves. References are given for texts and authors to help those who desire a more detailed understanding of the theoretical aspects.

The author appreciates the help and material given by Bill Stuehr of Induction Tooling and the material furnished by Robert Ruffini of Fluxtrol and George Welch of Ajax Magnathermic.

Contents

CHAPTER 1: Heat Treating of Metal 1
CHAPTER 2: Theory of Heating by Induction 5
CHAPTER 3: Induction Heat Treating Systems 19
CHAPTER 4: Induction Coils 43
CHAPTER 5: Heat Treating Basics 75
CHAPTER 6: Quenching 121
CHAPTER 7: Tempering 137
CHAPTER 8: Cleaning and Rust Protection 149
CHAPTER 9: Decarburization and Defects 151
CHAPTER 10: Applications of Induction Heat Treatment ...165
CHAPTER 11: Induction Heat Treating Process Analysis ...183
CHAPTER 12: Standards and Inspection 215
CHAPTER 13: Nonconforming Product and Process Problems ..235
CHAPTER 14: Quality Control 249
CHAPTER 15: Maintenance 261
APPENDIX 1: Metallurgical Definitions for Induction Heat
    Treating 267
APPENDIX 2: Scan Hardening 279
APPENDIX 3: Induction Coil Design and Fabrication 283
APPENDIX 4: Quench System Design 299
APPENDIX 5: Induction Tempering 307
APPENDIX 6: Tempering Curves 311
APPENDIX 7: Hardenability Curves 315

CHAPTER **1**

Heat Treating of Metal

History

THE EGYPTIANS are believed to have worked copper for centuries before 3500 B.C. A piece of heat treated steel was found in one of the pyramids, and it is thought to date from 3000 B.C. Early metal workers found certain metals and ores could be refined, processed, and made into tools and weapons, but it was not until the Iron Age and the Hittites that metallurgical processes were developed that would consistently produce strong steel weapons. Although the art of metallurgy developed as early smiths found that heating and cooling iron in different ways could make iron either softer or harder, the metallurgical theory lagged behind until relatively modern times. In 1864 Henry Clifton Sorby first used a microscope to study metals. This was followed by Albert Sauveur who tried to convince American steelmakers that something practical was to be gained from microscopic examination. However, it has only been since about 1930, when x-ray diffraction with wave mechanics was applied to metals, that the science of metallurgy was born.

The first induction phenomenon was observed by Michael Faraday in the middle 1800s when the effect that caused the heating of transformer and motor windings was considered to be undesirable. The first constructive use of induction occurred in 1916 when it was used to melt metals. Induction heat treating came into prominence in the 1930s, when high-frequency motor generator sets were developed and used for the induction hardening of crankshaft journals and bearings. In 1938, Caterpillar installed a power supply for induction-hardening track links, and by 1943 they had 16 induction-hardening units in production (Ref 1). In 1941 Vaugn, Farlow, and Meyer presented a paper titled "Metallurgical Control of Induction Hardening" at the convention of the American Society for Metals, which provided proof that alloy elements such as nickel and chromium were wholly unnecessary for maximum surface hardness and that carbon steels could be used in place of alloy steels. Caterpillar subsequently purchased a 500 kW, 9.6 kHz motor generator set for induction hardening their final drive gear with a 642 mm (25.7 in.) diameter by a 125 mm (5 in.) wide face. In an article in the July 1943 issue of *Metal Progress*, the Caterpillar process for contour hardening this gear was presented. Figure 1.1 shows the contour pattern produced at that time by Caterpillar. Caterpillar must be considered the early pioneer in the contour hardening of gear teeth.

Progress in research in metallurgical principles of induction hardening continued, and at the 26th annual meeting of the American Society for Metals in 1944, D.L. Marten and F. E. Wiley presented a paper that reported investigation of temperature, composition, and previous structure upon induction-hardening characteristics of plain carbon steel (Ref 2). The basic metallurgical theory as presented at that time is still being taught today.

In 1946, Edwin Cady listed the basic types of induction equipment (Ref 3) with frequencies ranging from 25 Hz to 50 MHz:

- Electronic circuit (vacuum tube, 300 to 530 kHz, and greater than 1 MHz)
- Spark gap (15 kHz to 60 kHz and 125 kHz to 450 kHz)
- Rotary converter (motor generator, 1 kHz to 10 kHz)
- Mercury arc (400 Hz to 3 kHz)
- Standard power cycle (line frequency of 60 Hz, or 25 Hz as generated by some steel mills)

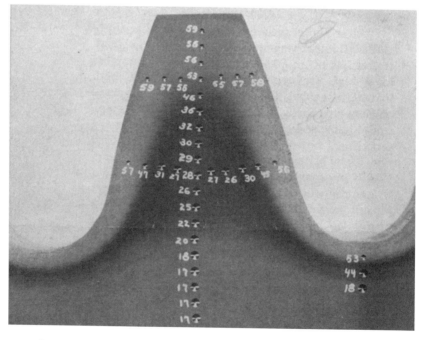

Fig. 1.1 Hardness survey (Rockwell C scale) of hardened tooth, sectioned on center. Magnified 2¾ diameters. Source: Ref 1

The use and development of induction heat treating practices continued to grow after World War II, and output transformers were developed to help the power supplies and load match when using low turn work coils. From the 1940s through the 1950s, the use and application of large motor generators and radio frequency (RF) oscillator induction power supplies continued. In the 1960s solid-state power supplies were invented for the conversion of line frequency into medium frequency induction heating. Because of their higher efficiency and increased versatility as the reliability of the solid-state power supplies increased, solid-state started to replace motor generator sets in the 1970s. The continued development trend of solid-state power supplies has been into higher frequencies as the solid-state devices have continued to increase in current-carrying and voltage-blocking characteristics. Today, while there is still a market for RF oscillators, most of the induction heating equipment sold is solid-state. Solid-state has made even the RF oscillators more efficient through replacement of the tube rectifiers by solid state diodes. If the past is used as a basis for projection into the future, the probability is that transistors will continue to improve and at some time will power all RF power supplies.

Advantages

Induction heating has the ability to rapidly heat specific areas of a part, such as the teeth of a gear or the bearing area of a shaft. Not only can superior mechanical properties be produced in such an area, but also the entire part does not have to be heated as is done with furnace heat treating. Significant benefits are produced such as:

- *Superior mechanical properties:* A hard case and a soft core provide a good blend of strength and toughness not attainable with furnace through heating. Furthermore, because the hardness of as-quenched steels depends only on carbon content, carbon steels can be used instead of alloy steels for most applications. Induction hardened tractor axles have a significant increase in bending fatigue over axles that are conventionally furnace hardened. Axles and shafts are also induction case hardened to produce high torsional strength, and many parts such as gears are selectively induction hardened to provide wear resistance on the gear teeth.
- *Lower manufacturing costs:* Total energy costs can be reduced because the entire part does not have to be heated. The costs of other processes that are necessary for furnace-hardened parts are reduced because the lower distortion produces the need for less grinding and finishing for final net shape. Straightening can sometimes be eliminated.

- *Manufacturing compatibility:* Induction heat treating systems can be automated for high production requirements and can be incorporated into manufacturing cells. Floor space requirements are reduced, and the workplace operating environment is improved.

The development and use of solid-state power supplies for induction heating continues today in all frequency ranges. There are many different applications for induction heating outside the heat treating area, but only heat treating of steel will be discussed in this book.

There are many other historic specific terms used for induction heating power supplies such as converters, inverters, motor generators, vacuum tube oscillator, spark gap generator, frequency tripler, and so on. These frequency converters change the 50/60 Hz line frequency to higher frequency. This book is concerned with the application of power supplies that are most commonly used in induction heat treating practice that change or convert three-phase, 50/60 Hz line frequency into single-phase high frequencies above 3 kHz. While there are installations and systems that use frequencies below 3 kHz for heat treating, they relate to a smaller number of specific installations rather than wide and varied commercial use and deal with dedicated, through-heating type applications such as the heat treating of pipe. Induction is also used widely for forging, melting, and a good number of individual applications. For purposes of this book, the terms *induction heater* and *power supply* will essentially mean the same thing: An induction heater is a power supply that produces the high frequency for induction heating.

REFERENCES

1. G.C. Riegel, Casehardening Large Gears with High Frequency, *Met. Prog.*, July 1943, p 82
2. D.L. Marten and F.E. Wiley, *Induction Hardening of Plain Carbon Steels: A Study of the Effect of Temperature, Composition, and Prior Structure on the Harden and Structure after Hardening*, Vol 34, Transactions of the ASM, 1945, p 351–404
3. E. Cady, Induction Heating, *Materials and Methods*, Aug 1946, p 401

CHAPTER 2

Theory of Heating by Induction

INDUCTION HEATING was first noted when it was found that heat was produced in transformer and motor windings, as mentioned in the Chapter "Heat Treating of Metal" in this book. Accordingly, the theory of induction heating was studied so that motors and transformers could be built for maximum efficiency by minimizing heating losses. The development of high-frequency induction power supplies provided a means of using induction heating for surface hardening. The early use of induction involved trial and error with built-up personal knowledge of specific applications, but a lack of understanding of the basic principles. Throughout the years the understanding of the basic principles has been expanded, extending currently into computer modeling of heating applications and processes. Knowledge of these basic theories of induction heating helps to understand the application of induction heating as applied to induction heat treating. Induction heating occurs due to electromagnetic force fields producing an electrical current in a part. The parts heat due to the resistance to the flow of this electric current.

Resistance

All metals conduct electricity, while offering resistance to the flow of this electricity. The resistance to this flow of current causes losses in power that show up in the form of heat. This is because, according to the law of conservation of energy, energy is transformed from one form to another—not lost. The losses produced by resistance are based upon the basic electrical formula: $P = i^2R$, where i is the amount of current, and R is the resistance. Because the amount of loss is proportional to the square of the current, doubling the current significantly increases the losses (or heat) produced. Some metals, such as silver and copper, have very low resistance and, consequently, are very good conductors. Silver is expensive and is not ordinarily used for electrical wire (although there were some induction heaters built in World War II that had silver wiring because of the copper shortage). Copper wires are used to carry electricity through power lines because of the low heat losses during transmission. Other metals, such as steel, have high resistance to an electric current, so that when an electric current is passed through steel, substantial heat is produced. The steel heating coil on top of an electric stove is an example of heating due to the resistance to the flow of the household, 60 Hz electric current. In a similar manner, the heat produced in a part in an induction coil is due to the electrical current circulating in the part.

Alternating Current and Electromagnetism

Induction heaters are used to provide alternating electric current to an electric coil (the induction coil). The induction coil becomes the electrical (heat) source that induces an electrical current into the metal part to be heated (called the workpiece). No contact is required between the workpiece and the induction coil as the heat source, and the heat is restricted to localized areas or surface zones immediately adjacent to the coil. This is because the alternating current (ac) in an induction coil has an invisible force field (elec-

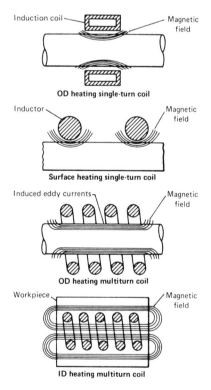

Fig. 2.1 Induction coil with electromagnetic field. OD, outside diameter; ID, inside diameter. Source: Ref 1

tromagnetic, or flux) around it. When the induction coil is placed next to or around a workpiece, the lines of force concentrate in the air gap between the coil and the workpiece. The induction coil actually functions as a transformer primary, with the workpiece to be heated becoming the transformer secondary. The force field surrounding the induction coil induces an equal and opposing electric current in the workpiece, with the workpiece then heating due to the resistance to the flow of this induced electric current. The rate of heating of the workpiece is dependent on the frequency of the induced current, the intensity of the induced current, the specific heat of the material, the magnetic permeability of the material, and the resistance of the material to the flow of current. Figure 2.1 shows an induction coil with the magnetic fields and induced currents produced by several coils. The induced currents are sometimes referred to as *eddy-currents*, with the highest intensity current being produced within the area of the intense magnetic fields.

Induction heat treating involves heating a workpiece from room temperature to a higher temperature, such as is required for induction tempering or induction austenitizing. The rates and efficiencies of heating depend upon the physical properties of the workpieces as they are being heated. These properties are temperature dependent, and the specific heat, magnetic permeability, and resistivity of metals change with temperature. Figure 2.2 shows the change in specific heat (ability to absorb heat) with temperature

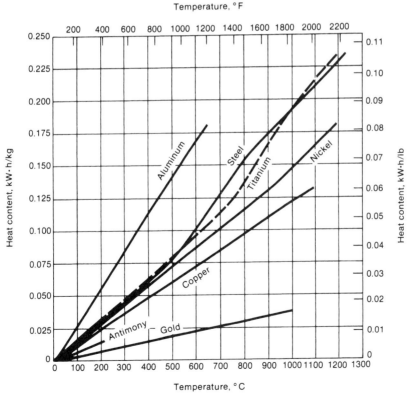

Fig. 2.2 Change in specific heat with temperature for materials. Source: Ref 2

for various materials. Steel has the ability to absorb more heat as temperature increases. This means that more energy is required to heat steel when it is hot than when it is cold. Table 2.1 shows the difference in resistivity at room temperature between copper and steel with steel showing about ten times higher resistance than copper. At 760 °C (1400 °F) steel exhibits an increase in resistivity of about ten times larger than when at room temperature. Finally, the magnetic permeability of steel is high at room temperature, but at the Curie temperature, just above 760 °C (1400 °F), steels become nonmagnetic with the effect that the permeability becomes the same as air.

Hysteresis

Hysteresis losses occur only in magnetic materials such as steel, nickel, and a few other metals. As magnetic parts are being heated, such as those made from carbon steels, by induction from room temperature, the alternating magnetic flux field causes the magnetic dipoles of the material to oscillate as the magnetic poles change their polar orientation every cycle. This oscillation is called *hysteresis*, and a minor amount of heat is produced due to the friction produced when the dipoles oscillate. When steels are heated above Curie temperature they become nonmagnetic, and hysteresis ceases. Because the steel is nonmagnetic, no reversal of dipoles can

Table 2.1 Resistivity of different metals

Material	\multicolumn{8}{c}{Approximate electrical resistivity, $\mu\Omega \cdot$ cm ($\mu\Omega \cdot$ in.), at temperature, °C (°F), of:}
	20 (68)	95 (200)	205 (400)	315 (600)	540 (1000)	760 (1400)	980 (1800)	1205 (2200)
Aluminum	2.8 (1.12)	...	...	6.9 (2.7)	10.4 (4.1)	...	...	...
Antimony	39.4 (15.5)	...	...	...	...	...	...	...
Beryllium	6.1 (2.47)	...	...	...	11.4 (4.5)	...	...	...
Brass(70Cu-30Zn)	6.3 (2.4)	...	...	...	...	...	...	...
Carbon	3353 (1320.0)	...	...	...	1828.8 (720.0)	...	...	...
Chromium	12.7 (5.0)	...	...	...	...	...	...	...
Copper	1.7 (0.68)	...	...	3.8 (1.5)	5.5 (2.15)	...	9.4 (3.7)	...
Gold	2.4 (0.95)	...	...	...	...	...	12.2 (4.8)	...
Iron	10.2 (4.0)	14.0 (5.5)	...	...	63.5 (25.0)	106.7 (42.0)	123.2 (48.5)	...
Lead	20.8 (8.2)	27.4 (10.8)	...	49.8 (19.6)	...	...	...	...
Magnesium	4.5 (1.76)	...	...	...	...	...	...	...
Manganese	185 (73.0)	...	...	...	...	...	...	...
Mercury	9.7 (3.8)	...	...	...	...	...	...	...
Molybdenum	5.3 (2.1)	...	...	...	...	...	...	33.0 (13.0)
Monel	44.2 (17.4)	...	...	...	...	...	...	...
Nichrome	108.0 (42.5)	...	...	114.3 (45.0)	...	114.3 (45.0)	...	...
Nickel	6.9 (2.7)	...	...	29.2 (11.5)	40.4 (15.9)	...	54.4 (21.4)	...
Platinum	9.9 (3.9)	...	...	...	...	...	...	...
Silver	1.59 (0.626)	...	...	...	...	6.7 (2.65)	...	...
Stainless steel, nonmagnetic	73.7 (29.0)	...	...	99.1 (39.0)	...	...	130.8 (51.5)	...
Stainless steel 410	62.2 (24.5)	...	...	...	101.6 (40.0)	...	127 (50.0)	...
Steel, low carbon	12.7 (5.0)	16.5 (6.5)	...	...	59.7 (23.5)	102 (40.0)	115.6 (45.5)	121.9 (48.0)
Steel, 1.0% C	18.8 (7.4)	22.9 (9.0)	...	...	69.9 (27.5)	108 (42.5)	121.9 (48.0)	127.0 (50.0)
Tin	11.4 (4.5)	...	20.3 (8.0)	...	...	...	...	...
Titanium	53.3 (21.0)	...	...	...	...	...	...	165.1 (65.0)
Tungsten	5.6 (2.2)	...	...	...	...	...	...	38.6 (15.2)
Uranium	32.0 (12.6)	...	...	...	...	...	...	...
Zirconium	40.6 (16.0)	...	...	...	...	...	...	...

Source: Ref 3

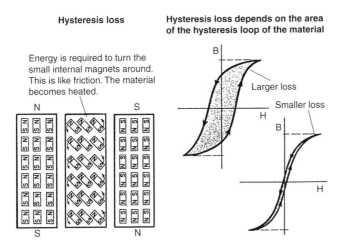

Fig. 2.3 Effect of hysteresis on heating rate. N, north; S, south; B, flux density in a ferromagnetic material; H, corresponding magnetic intensity. Source: Ref 4

occur. Figure 2.3 shows an illustration of hysteresis and the effect on the magnetic flux field strength. Figure 2.4, as represented by the line "ABCD," shows the Curie temperature for carbon steels.

Skin Effect and Reference Depth

Induction heating occurs when an electrical current (eddy current) is induced into a workpiece that is a poor conductor of electricity. For the induction heating process to be efficient and practical, certain relationships of the frequency of the electromagnetic field that produces the eddy currents, and the properties of the workpiece, must be satisfied. The basic nature of induction heating is that the eddy currents are produced on the outside of the workpiece in what is often referred to as "skin effect" heating. Because almost all of the heat is produced at the surface, the eddy currents flowing in a cylindrical workpiece will be most intense at the outer surface, while the currents at the center are negligible. The depth of heating depends on the frequency of the ac field, the electrical resistivity, and the relative magnetic permeability of the workpiece. For practical purposes of understanding, the skin heating effect (reference depth) is defined as the depth at which approximately 86% of the heating due to resistance of the current flow occurs. Figure 2.5 shows reference depths for various materials at different temperatures. The reference depths decrease with higher frequency and increase with higher temperature. The reference depth, as mentioned, becomes the theoretical minimum depth of heating that a given frequency will produce at a given power and workpiece temperature. The cross-sectional size of the workpiece being heated must be at four times the reference depth, or what appears to be current cancella-

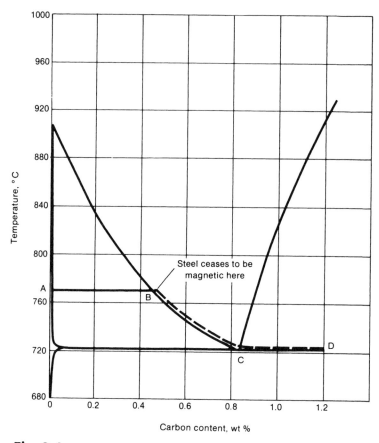

Fig. 2.4 Curie temperature for carbon steels. Source: Ref 2

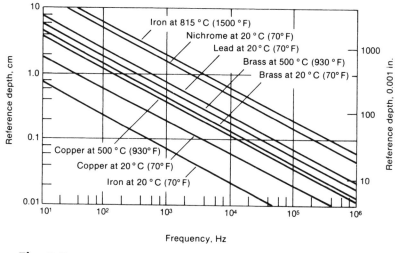

Fig. 2.5 Reference depth for various materials. Source: Ref 2

tion occurs. Figure 2.6 shows the critical frequency (or minimum frequency) for heating different bar diameters. Figure 2.7.a illustrates three examples of ratios of workpiece thickness *(a)* and reference depth of heating *(d)* with the respective current distribution. The dashed lines show exponential decay from either side, while the solid line gives the net current from the summation of the two dashed lines. As the workpiece thickness/reference depth of heating ratio decreases below four to one, the net current decreases. The net heating curves (Fig. 2.7.b) are obtained by squaring the net current density, demonstrating that when $a/d = 4$, the best surface heat distribution occurs.

For a fixed frequency, the reference depth varies with temperature because the resistivity of conductors varies with temperature. With magnetic steels the magnetic permeability varies with temperature, decreasing to a value of one (the same as free space) at the Curie temperature, at which steel becomes nonmagnetic. Because the reference depth increases when steel is heated over the Curie temperature, the a/d ratio of 4 when austenitizing must be based on the reference depth when the steel is at a temperature above the Curie. Figure 2.8 shows an illustration of the deeper depth of current penetration over the Curie. Because of these effects the reference depth of nonmagnetic materials may vary by a factor of two or three over a wide heating range, whereas for magnetic steels it can vary by a factor of 20. The net effect is that cold steel has a very shallow reference depth as compared to hot steel.

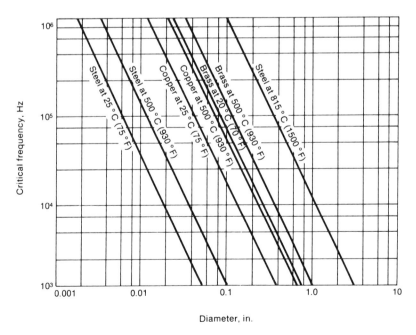

Fig. 2.6 Critical frequency for efficient heating of several materials. Source: Ref 2

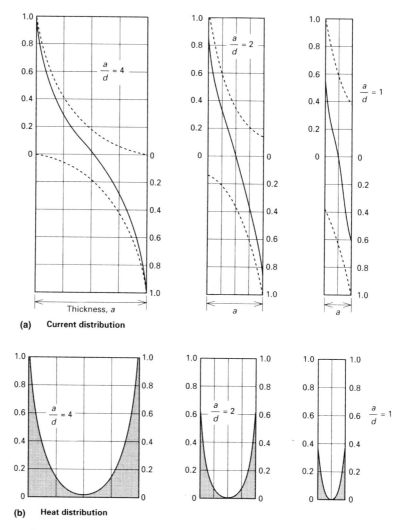

Fig. 2.7 (a) Ratios of object thickness, a, and reference depth, d. (b) Net heating curves. Source: Ref 5

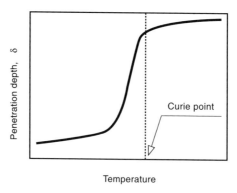

Fig. 2.8 Variation of current penetration depth through Curie. Source: Ref 6

Another very important result of the skin effect is evidenced in heating efficiency. Heating efficiency is the percentage of the energy put through the coil that is transferred to the workpiece by induction. As shown in Fig. 2.7, if the ratio of workpiece diameter to reference depth for a round bar drops below about 4 to 1, the heating efficiency drops. This ratio becomes what is defined for round bars as the critical frequency for heating. Figure 2.9 shows the critical frequency as a function of diameter for round bars and Figure 2.10 shows the efficiency of heating as a function of this critical frequency. Higher frequencies are needed to efficiently heat small bars, but as Figure 2.10 shows, once the critical frequency is reached, increasing the frequency has very little effect on relative efficiency.

For through heating, a frequency close to the critical frequency should be selected so that the workpiece will through heat faster. In contrast for case hardening, frequencies higher than the four-to-one ratio will be selected based on a combination of being both higher than the critical frequency and high enough to generate the desired skin effect heating for

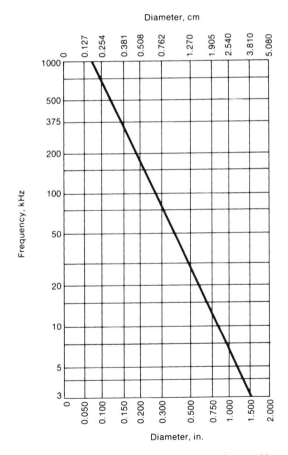

Fig. 2.9 Critical temperature as a function of diameter for round bars. Source: Ref 2

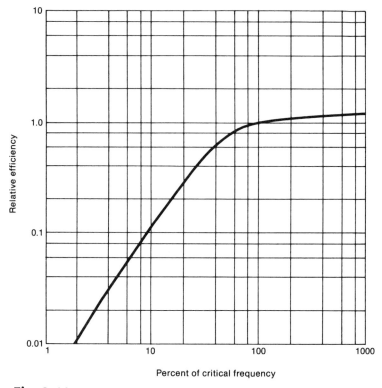

Fig. 2.10 Efficiency of heating as a function of frequency. Source: Ref 2

production of the specified case. Once the critical frequency is reached, the case depth requirements will help in frequency selection because lower frequencies have deeper reference depths, thereby producing deeper case depths. The Chapter "Induction Heat Treating Process Analysis" in this book discusses frequency selection.

Power Density

Selection of power is just as important as the selection of frequency. When case hardening is to be done, the short heat cycles that are necessary require higher power density (energy input per unit of surface) than through-heating applications. The power density at the induction coil is the metered output power divided by the amount of workpiece surface within the induction coil and is expressed in kW/cm^2 or $kW/in.^2$. Power density requirements, as shown subsequently, can be used to rate the power requirements for an application.

Power requirements are related to the amount of energy required to heat a workpiece and to the induction heating system power losses. The energy or heat content required to heat the workpiece can be calculated when the material, its specific heat, and the effective weight of material to be heated per hour are known.

The value kWh is (lbs/hour × specific heat × temperature rise)/ 3413, where the pounds per hour relate to (3,600 seconds × part weight)/actual heat cycle. Heat input required to heat a specific workpiece represents only the energy or power that needs to be induced from the coil into the workpiece. Other system losses such as coil losses, transmission losses, conversion losses, and the ability to load match for required output power determine the power supply rating. Table 2.2 shows the typical system losses for a typical solid-state power supply system and a radio frequency (RF) vacuum tube system. Through calculation of the heat content requirements of the workpiece at the coil, and with the system losses known, the power requirements for heating can be determined. If a heat content of 25 kWh is needed by the workpiece and system losses are 50%, then a minimum output power rating of 50 kW is needed from the power supply. As discussed in the Chapter, "Induction Heat Treating Systems" in this book, the ability of a power supply to produce rated output power depends on the ability to load match the power supply to the induction coil. When in doubt it is advisable, to use power supplies with higher output power ratings than needed. The Chapter "Induction Heat Treating Process Analysis" in this book deals with the determination of power supply ratings from tables that have been developed through calculation and correlation with surface power densities and for through heating. From calculation of the surface of the area to be heated in relationship to power, power density curves define power supply ratings. For instance if a power density of 1.55 kW/cm^2 (10 kW/in.^{2}) is needed and the area to be heated is 5 square inches, then the power required is 50 kW. In general, obtaining higher production rates for specific case depths for surface hardening requires higher power densities.

Through heating systems can be defined from knowing the cross-sectional size and the weight of steel to be heated per hour. Lower power densities and frequencies are used for through heating because the workpieces need to have the heat soak and penetrate to the core. Higher productivity is obtained from using more power heating and from heating more workpiece area at the same time, thereby not increasing the power density.

Higher power densities provide the ability to heat surfaces more rapidly. However, there may be limitations to the amount of power that an individual induction coil can handle. Induction coil design is discussed in

Table 2.2 Typical system losses for different induction power supplies

Power supply	Frequency	Terminal efficiency, %	Output transformer efficiency, %	Coil efficiency, %	System efficiency, %
Solid state	10 kHz	90	75	75	51
Radio frequency	450 kHz	65	60	85	33

Source: Ref 7

the Appendix "Induction Coil Design and Fabrication" in this book. The amount of cooling water a given coil can carry defines the amount of power it can carry. Also, the fact that the power supply has a more than adequate power rating does not necessarily mean that the power can be induced into the workpiece. Only 125 kW of a 250 kW power supply may load into a given coil and part. Load matching and tuning are discussed in the Chapter "Induction Heat Treating Systems" in this book. Tables of power densities are listed in that Chapter as well, and tables of scanning speeds are listed in the Appendix "Scan Hardening" in this book.

Conduction of Heat

The primary mechanism of heat flow to the interior of a workpiece comes from the conduction of the heat first produced by the eddy currents on the surface. Hysteresis produces a secondary effect and is a small, second producer of heat. Losses due to hysteresis are usually ignored in the heat content calculations for induction processing because of the minor effect. As previously discussed, the heating rate is mainly affected by:

- Field strength of the magnetic flux field
- Coupling of the induction coil to the workpiece
- The electrical and magnetic properties of the material being heated

The desired rate of heating varies with the application. Mathematical analysis of heat transfer can be quite complex because of the interreaction of the intense heat produced due to the eddy-current heating of the surface, the rate of heat transfer toward the core, and the fact that the electrical, thermal, and metallurgical properties of most materials exhibit a strong dependence on temperature and vary during the heating cycle.

One way to view the heat propagation is as a wave format. Using carbon steel as an example, high heat is first produced in the reference depth of the magnetic material. When the Curie temperature is passed, the depth being heated increases in a wavelike motion until the reference depth of the nonmagnetic material is reached. From that point most of the heat transfer occurs by the inward conduction of heat. Figure 2.11 shows the temperature profile of heat for 25 kHz on a 50 mm (2 in.) cylinder. At 0.4 s the surface being heated is being heated under the Curie and is nonmagnetic. At 1.15 s the surface is nonmagnetic, and the surface is heating deeper with minor temperature migration into the core. At 6.5 s the austenitic portion of the surface has increased to about 0.200 in. deep, and there has been minor temperature migration to the core. It should be noted

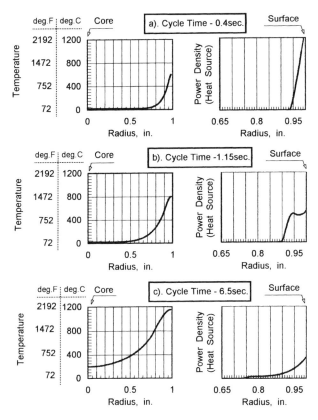

Fig. 2.11 Interrelationship among heating time, power density, and hardened depth. Source: Ref 6

that at this point in order to accomplish the heat transfer, the surface temperature has increased to 1200 °C (2192 °F). Heat transfer is a matter of temperature differential and the thermal conduction rate of the workpiece. When the material is austenitic, higher temperature differential is necessary at the surface to increase the speed of heat conduction and transfer toward the core. As soon as the power is turned off, the increase in temperature seems to stop at the edge of the austenitized area. When heat treating parts, the case depth produced during heating does not appear to penetrate any deeper while the part is being quenched (except in the circumstance of small diameter parts). During case hardening, cooling occurs from both the outer surface on which the quenchant is being applied, and the core towards which the heat is migrating. Holes under the case can affect both the eddy-current distribution and heat transfer. This is discussed in the Chapter "Induction Heat Treating Process Analysis" in this book.

The important thing to remember in frequency selection is that smaller parts need higher frequencies, and that the lowest frequency that will heat a given cross section is usually the most economical.

REFERENCES

1. *Heat Treating*, Vol 4, *Metals Handbook*, 9th ed., American Society for Metals, 1981
2. S.L. Semiatin and D.E. Stutz, *Induction Heat Treating of Steel*, American Society for Metals, 1986
3. S.L. Semiatin and S. Zinn, *Induction Heat Treating*, ASM International, 1988
4. C.A. Tudbury, *Basics of Induction Heating*, John F. Rider, Inc., New Rochelle, NY, 1960
5. *Heat Treating*, Vol 4, *ASM Handbook*, ASM International, 1991
6. V. Rudnev et al., *SteelHeat Treatment Handbook*, Marcel Dekker, Inc., 1997
7. R.E. Haimbaugh, Induction Heat Treating Corp., personal research

CHAPTER **3**

Induction Heat Treating Systems

THE ELEMENTS that make up an induction heat treating system are shown in Fig. 3.1. These elements are the cooling water for electrical cooling, power supply, heat station, work-handling fixture, controls, work coil, and quench system. Depending on the power supply and processing requirements, these elements may be furnished either separately or unitized. This Chapter begins with a discussion of the types of power supplies and then moves into discussion of the system elements and their influence on and requirements for induction heat treating system design.

Types of Power Supplies

As mentioned earlier, there are many different types of induction heating power sources, ranging from line frequency coils, to heat for shrink fitting, to many different types of induction power supplies. Most induction power supplies sold for heat treating today are either some type of solid state or oscillator (vacuum) tube. Many types and models of induction power supplies are made to meet the diverse requirements for different frequencies and output power requirements for induction heat treating. Regardless of the electronic technology, the power supplies perform a common function. Figure 3.2 shows a block diagram of modern high-frequency power supplies performing line frequency conversion into high frequency. The power supplies are basically frequency changers that change the 60 Hz (U.S.), three-phase current furnished by the electric utility into a higher-frequency, single-phase current for induction heating. These power supplies are often referred to as converters, inverters, or oscillators, depending on the circuits and electronic devices used, with many possible combinations of conversion techniques. Because of the complexity of the circuit and controls, it can be difficult to distinguish the difference in operating characteristics between the power supplies sold by various manufacturers because the manufacturers' claims are technically difficult to verify. There is no current standard for rating the output power of solid-state power supplies except for line voltage and current requirements (which do not guarantee rated output power for specific applications). RF oscillator tube power supplies are supposed to be rated by water load at the coil.

Solid-state power supplies convert the line alternating voltage (ac) to produce single-phase, direct-current (dc) voltage. Inversion is then accomplished through using thyristors (silicon controlled rectifiers, or SCRs), or transistors such as isolated gate bipolar transistors (IGBTs) or metal-silicon-dioxide field-effect transistors (MOS FETs), to produce dc pulses that are then made sinusoidal to form high frequency, ac. (Some current source power supplies do this in one step.) Radio frequency (RF) (oscillator or vacuum tube) power supplies use a transformer to change the input voltage to high voltage before conversion to dc. The oscillator tube is used

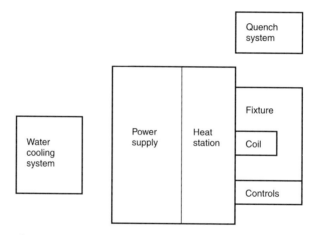

Fig. 3.1 Elements of an induction heat treating system. Source: Ref 1

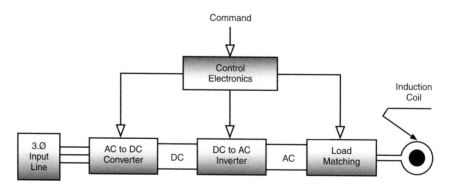

Fig. 3.2 Induction heat treat power supply basic diagram. ac, alternating current; dc, direct current. Source: Ref 2

to produce dc pulses that are likewise changed into the high-frequency, ac current. The higher output voltage of the RF tube power supplies is one of their more distinguishable features.

The frequency furnished by the power supply is critical to the intended induction heat treating process because of the relationships between the size of the part that is being heated and the depth of heating of the frequency being used. In the discussion of reference depth in the Chapter "Theory of Heating by Induction," in this book, it was mentioned that the higher the frequency, the shallower the hardened depth and the better the efficiency at the coil. For the purposes of this book, induction heat treating practice will deal with most commonly used frequencies above 3 kHz, although there are a few applications for lower frequencies. If the use of a frequency below 3 kHz is needed for a particular application, the theory of frequency use and power requirements is used with the lower frequencies in similar fashion as with the higher frequencies. One distinguishable feature below 3 kHz is that the coils have more mechanical vibration, so rigidity in the coil design can be very important.

Figure 3.3 shows the use of SCRs, transistors (IGBTs and MOS FETs), and vacuum tube oscillators currently in use. At the lower end of the frequency range, up to 10 kHz, SCRs are in wide use for the switching devices because of device cost versus current carrying capabilities. In the medium frequencies IGBT transistors are used, and in the higher frequencies MOS FET transistors are used. In the future, as the current-carrying ability of transistors is increased and the cost is decreased, transistors are expected to come into wider use over the full frequency range. Solid-state power supplies and the RF oscillator tube power supplies have considerable differences in efficiency, as shown in Table 3.1. The lower frequency, solid-state power supplies are more efficient in energy conversion. The RF oscillator tube has a filament that consumes energy being heated all the time, and the switching losses in oscillator tubes are high.

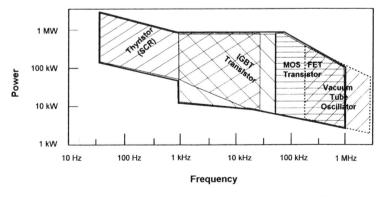

Fig. 3.3 Modern inverter power types for heat treating. SCR, silicon controlled rectifier; IGBT, isolated gate bipolar transistor; MOS FET, metal-silicon-dioxide field-effect transistor. Source: Ref 3

The two main types of inverters used in solid-state are voltage-fed and current-fed. These can be further subdivided by the dc source as shown in Fig. 3.4. Versions of these different inverters are sold by the various induction heating equipment manufacturers. Reference 3 goes into the theoreti-

Table 3.1 Comparative efficiencies of various power sources

Power source	Frequency	Terminal efficiency, %	Coil efficiency, %	System efficiency, %
Supply system	50 to 60 Hz	93 to 97	50 to 90	45 to 85
Frequency multiplier	50 to 180 Hz	85 to 90	50 to 90	40 to 80
	150 to 540 Hz	93 to 95	60 to 92	55 to 85
Motor-generator	1 kHz	85 to 90	67 to 93	55 to 80
	3 kHz	83 to 88	70 to 95	55 to 80
	10 kHz	75 to 83	75 to 96	55 to 80
Static inverter	500 Hz	92 to 96	60 to 92	55 to 85
	1 kHz	91 to 95	70 to 93	60 to 85
	3 kHz	90 to 93	70 to 95	60 to 85
	10 kHz	87 to 90	76 to 96	60 to 85
Radio-frequency generator	200 to 500 kHz	55 to 65	92 to 96	50 to 60

Source: Ref 4

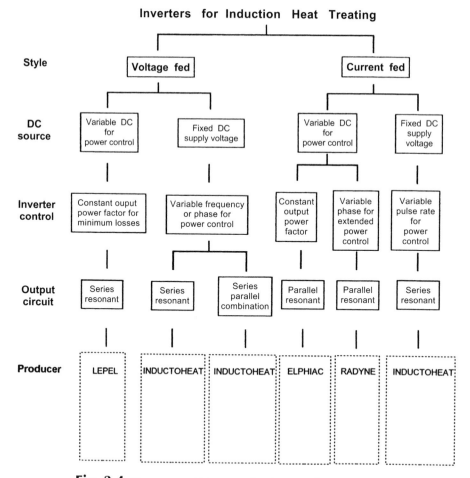

Fig. 3.4 Direct current (dc) source of main types of inverters. Source: Ref 3

cal explanation of the electrical circuits for each source. Figure 3.5 shows the electric features of a bridge inverter. Current source inverters operate better with what is termed *high Q* loads, which essentially are loads that are more loosely coupled or tuned. However, as discussed later in this Chapter, the main factor that needs to be examined when selecting a power supply is the ability of the power supply to produce full power at the rated output power level over the band of impedance matching needed by the user.

Solid-State Advantages. Solid-state power supplies are preferred when the workpieces are large enough to permit cost-efficient frequency selection. High power units are less expensive and smaller in size than oscillator tube units, while having higher efficiency in conversion from line frequency to terminal output. Solid-state power supplies require no warm up, and they have a high degree of reliability. Finally, solid-state power supplies inherently have better power regulation with the ability to produce full power over an entire heating cycle. At the higher frequencies, such as above 50 kHz, the smaller MOS FET transistors are used. Higher frequencies cause more switching losses, resulting in reduction of the output power rating. With the higher frequencies, such as above 300 kHz, vacuum tube oscillators are still widely used.

Oscillator tube units operate in the 200 kHz up through 2 MHz frequency range and tend to have higher cost per kW of power sold. Older power supplies used rectifier tubes to complete the rectification to dc, while modern units use solid-state diodes. (The only tube in a modern power supply is the oscillator tube.) The output power of an RF oscillator decreases when magnetic steel parts are heated through the Curie temperature, so it is harder to maintain full power output. However, RF power supplies have been around for many years and have more versatility in impedance matching and tuning than solid-state power supplies. Radio frequency units are easy to tune, and when there is a component failure, they are easy to troubleshoot. Radio frequency tube power supplies have been in wide use for 50 years and have a good history of operation. Although oscillator tubes have 1,000 h warranties, tube life up to 25,000 h or more is not unusual.

Bridge Inverter Features	
Voltage Fed	Current Fed
DC Filter Capacitor	DC Inductor
Square Wave Voltage	Sine Wave Voltage
Sine Wave Current	Square Wave Current
Series Resonant Output	Parallel Resonant Output
Load Current = Output I.	Load Voltage = Output V.
Voltage x "Q"	Current x "Q"
Best For Low "Q" Loads	Best For High "Q" Loads

Fig. 3.5 Electric features of bridge inverter. dc, direct current; high "Q" loads, loads more loosely coupled or tuned. Source: Ref 3

While operating cost as determined by power supply cost and conversion efficiency to the workpiece is a consideration, other factors must also be reviewed. Once the desired frequency and power requirements have been calculated for a power supply, factors such as the recommended line utility requirements, cooling water requirements, system losses in the heat station and coil, power supply controls and regulation, and window for full power regulation, should be reviewed.

Line Utility Requirements

The input kilovolt-amps (kVA) and line voltage tolerance of the power supply should be reviewed. Local utilities are required to furnish a given range of line voltages. However, line voltage variation during a workday can produce output power fluctuations that may be a problem. Also, solid-state devices such as diodes and SCRs produce line harmonics back into the power line that can potentially affect other solid-state controls and equipment.

Cooling Water Requirements

Cooling water is required because all of the electrical energy consumed is essentially transferred in the form of heat to the cooling water. Except for small radiation losses and residual heat left after quenching, even the heat produced in the workpiece is absorbed by the quenchant, which in turn is cooled through a heat exchanger. If the heat input into the workpiece is 50 kW in a 100 kW induction heating system, then the other 50%, or 50 kW, represents system loss being absorbed by the cooling water. Thus at full power, a 100 kW system is transmitting 100 kW into cooling water.

Water quality is extremely important in induction heat treating systems. A review of the maintenance problems of induction heating systems shows poor water quality to be the leading cause of maintenance problems. There are six sources of problems in water:

- *pH level:* Cooling circuits consist mostly of brass and copper fittings and pipe. A range of 6 to 8 appears to be the best range for good results. Below 6 will produce corrosion of brass, and above 8 scale will form.
- *Suspended solids:* If allowed to accumulate, the suspended solids will tend to clog the small water passages in transformers and capacitors. Suspended solids tend to settle at the low points of cooling passages when the pumps are turned off, leading to eventual clogging.
- *Dissolved gases:* These are not generally a problem unless there is open air turbulence. Dissolved carbon dioxide can result in high corrosion rates by lowering the pH. Dissolved oxygen will increase elec-

trolytic corrosion rates and can be a problem when pressurized systems are used without dissolved gas removers.
- *Biological contaminants:* While the systems that contain copper do not usually have a problem with biological contamination, it is possible for algae or microorganisms to "infect" a cooling system. The contamination can result in corrosion, plugging, and reduced heat transfer. Professional help should be used for assistance in dealing with suspected biological problems.
- *Dissolved solids:* These are the most significant factor in cooling electrical equipment. Total dissolved solids directly affect the conductivity of the cooling water. When the conductivity is too high, electrolysis will occur in the cooling passages that have dc voltage potential. Different original equipment manufacturers (OEMs) have different methods of dealing with dissolved solids according to the individual requirements of their power supply, including the use of sacrificial targets and calculated cooling hose lengths. The maximum dissolved solids may be stated, such as 250 ppm. However, because the content of dissolved solids can be determined through the water conductivity, the recommendation of a conductivity such as 400 μS/cm maximum may be stated. Conductivity ranges will be covered when discussing closed water recirculation systems later in this Chapter.
- *Galvanic corrosion:* This is another cause of fitting deterioration and water contamination. Care must be taken in the connection of systems not to place dissimilar metals in contact with each other without the use of a galvanic fitting. For example, if the heat exchanger has copper pipe but the inlet water connection is made with steel pipe, galvanic fittings should be used.

The different elements of the induction heating system have different cooling water requirements. Thus the cooling water needed to cool the electrical components in the power supply, heat station, induction coil, and quench system can have different specifications. This is particularly important in large power systems so that cooling water costs can be minimized. Where large quantities of cooling water are needed, lower standards can be used to cool elements such as the quench system. The power supply OEMs give clear definitions of the quality, quantity, and maximum temperature of cooling water needed. While there has been a tendency to relax standards in some of the transistorized power supplies, over the long term high quality water is recommended for continued equipment operation.

Available cooling water sources on the line or input side include well water, river water, and reservoir or lake water, feeding into various types of open and closed systems. In the circumstances where a natural source of cooling water is not available, refrigerated cooling packages can be used. When a natural source of cooling water is used for cooling the power supply and heat station, whatever the source, the water should be supplied at a temperature above the dew point to prevent excessive condensation. Open water recirculating systems that add line water as needed to maintain the recirculated temperature work quite well. The excess water passes through an overflow to the drain for other manufacturing processes. Most natural water sources that can be legally used do not have high enough purity for use in cooling the power supplies, heat stations, and work coils. Engineers have tried many times to reduce the cost of power supplies by reducing water purity requirements, only to be proven wrong after the equipment is out of warranty. Purity and conductivity of the water for the power supply and heat station have historically proven important to reliable, long-term operation. The OEM specifications for water quality, flow, and minimum inlet temperature are the minimum specifications that should be met. It is the author's opinion that quenchants should not be used to cool power supplies and heat stations.

Different induction system configurations of power supply, heat station, and coil can have different inlet water pressure, water flow, and maximum water inlet requirements. From these requirements systems can be designed from the available line water that will provide the necessary cooling. Most available water supplies do not meet the water purity criteria; therefore, the line water is used to cool a separate system of high-purity cooling water that is recirculated through the induction system, plus any other separate systems that need cooling, such as the quench systems.

Figure 3.6 shows the preferred and most common type of cooling system used in cooling induction equipment. The system has a closed loop of

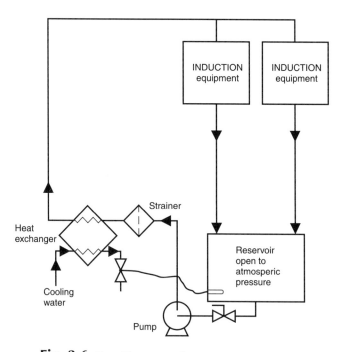

Fig. 3.6 Closed loop recirculation system. Source: Ref 5

cooling water running through all of the induction system and returning to an open tank. The system is nonpressurized, with the tank open to the atmosphere. By running the return lines into the tank below the water line, dissolved gases are minimized. This system should have all nonferrous plumbing and fittings. The closed side of the system recirculates high-purity cooling water at the correct temperature and flow through the induction system, with the cooling of the recirculated water accomplished in a heat exchanger. The open side of the system uses line water obtained from natural sources or other sources such as cooling towers or chillers.

Figure 3.7 shows the use of a common cooling system that uses an evaporative water tower to provide cooling for the line water side, adding make-up water as needed due to evaporation losses. These systems are most efficient when the summertime dew point on hot muggy days does not exceed the maximum temperature required on the line water side. (Above the dew point, water will not evaporate.) When this happens, other sources of cooling need to be used, such as adding make-up water or using refrigerated coolers for supplemental cooling. The line water sources may need to have biocides added to prevent biological contamination, and filters may be needed to keep out solids such as leaves and wind-blown debris. In the wintertime, antifreeze may be needed to prevent freezing on the tower side.

Where high-purity water is needed the internal closed systems of recirculated water commonly use distilled water or deionized water. When distilled water is used, the water should be changed on a periodic basis, such as every six months. New deionized water has a conductivity of about

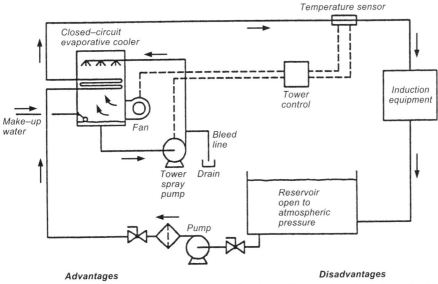

Fig. 3.7 Closed loop recirculation system with tower. Source: Ref 5

20 µS/cm and when first charged will dissolve some of the metallic components in the cooling system. The conductivity will increase very rapidly to about 180 to 200 µS/cm. After that, the conductivity will slowly increase to the 400 µS/cm level, at which point the water should be changed. Where deionized water is used with a closed loop deionizing system, it is recommended that the water be maintained at a specified conductive range such as 50 to 100 µS/cm. Because the water is in a closed system, biocides may be needed, and filters are recommended. Quench system design will be discussed in the Appendix "Quench System Design," in this book. Also, if the closed system needs antifreeze because of low plant ambient temperature, commercial antifreezes (as are used with cars) must not be used. They will ionize and break down under voltage. Only pure ethylene or propylene glycol should be used.

Heat stations contain the transformers and capacitors, as needed by a given power supply, to help match the power supply voltage or impedance to the coil. Heat stations may have tank circuits with the tuning capacitors either in parallel or in series with the induction coil. These components can be in the same enclosure as the power supply components, or they may be in a separate enclosure. High-frequency transformers tend to be inefficient, producing system power losses running from 10 to 40% of the output of the power supply. When the heat stations are furnished separately from the power supply, proper system design requires that the conductors carrying the high frequency be chosen and installed according to the instructions of the manufacturer of the equipment. Remote location, even exceeding a meter or several feet, can produce system losses that reduce the output power available at the coil. Likewise, coil design, as discussed in the Appendix "Induction Coil Design and Fabrication" in this book, is important in optimizing the amount of power available to be induced into a workpiece. Proper coil design reduces the power loss in the coil, enabling more power to be produced in the workpiece.

Power Supply Regulation

The ability to produce the same amount of power for each part during the heating cycle is important. The power supplies need to have regulation for two reasons. The first is so that the power supply has constant output if there is a change of the line voltage during operation (not an unusual circumstance). The manufacturers normally rate their power supply over the range of input voltage for which constant output can be produced. However, if the power supplies are being run at full power, regulation may not be achieved over large voltage swings, such as the line voltage ranging from 480 V down to 430 V in the same day. Radio frequency oscillator power supplies with front-end SCR power controllers are supposed to regulate, but they too may fail to handle large swings in line voltage. The RF oscillator power supplies may have other methods of regulation to help with small input voltage fluctuations, such as filament voltage regulation. Radio frequency power supplies do not regulate the output through closed loop feedback for frequency adjustment such as solid-state power supplies. Plate amperage, the metered indication of output current of the oscillator tube, decreases when the temperature of the workpiece is passing through the Curie temperature. Solid-state power supplies regulate to constant output through closed-loop controls to hold constant power, constant voltage, or constant current. The firing rate of the semiconductors is changed to cause a frequency shift so that the constant output is maintained. Each method of regulation of solid-state has areas of application. When the controls are set to hold constant power, there is no drop of power during heating. Voltage and current regulation tend to the output power better when there are wide variations in impedance during the heating cycle, such as when a scanner passes over different diameters of a workpiece when heating. Regulation of the power supply is important so that the total power produced into the workpiece during each heat cycle is the same during low-line voltage conditions or when the line voltage fluctuates. When purchasing a power supply, the buyer should be certain that the power supply regulates the output power for consistent output with normal-line voltage fluctuations that exist in the buyer's plant not only throughout the day, but also throughout the year.

Controls can be separated into three different groupings: the metering, the logic portion, and the operating control. The metering of the power supply provides the information not only for initial tuning, but also for monitoring the power supply performance during operation. Use of the meters is discussed in this Chapter under "Load Matching". Different types of power supplies have different metering requirements, and there are some variations among the different manufacturers as to what meter output is furnished. Factors to be considered are the types of metering furnished and the location of the metering (can you see the meters from the control, "heat on," "heat off" location).

The logic types of control are internal in the power supply. These include the automatic controls within the power supply such as various limit circuits, overload circuits, trip circuits, status indications, and power regulation circuits. Limit circuits serve to protect the power supply against conditions such as high frequency, high voltage, or high current. The overload circuits protect against detrimental or harmful conditions to the power supply such as short circuits, open circuits, and arcs. The internal control circuits also provide power regulation, such as constant power through both the heating cycles on solid-state equipment, to maintain a constant output. Also included in this type of control are any special circuits that tie in with remote-control feedback for automatic power control. Examples of control feedback circuits are those for which radiation pyrometers or linear speed feedback controllers provide signals that are used to regulate the output power back to the power supply.

The third type of control is the operating control, such as output power adjustment, the control that turns the power on and off. If 100% power is being produced and the workpiece is being heated too fast, adjustment of the power control will reduce power. Different types of power supplies do this differently, by either reducing the firing rate of the semiconductors or using front-end solid-state controllers to reduce the input voltage before rectification. It should be noted that it is possible to purchase some power supplies, such as RF tube sets, that do not have a stepless type of power control.

Turn-On Power Ramp. The first power supplies used mechanical relay-type contactors that provided almost instantaneous turn-on after closure. Today all power supplies that are solid-state or with SCR front ends have a ramp in which the power is increased from zero to full power to start the heat cycle. Slower ramps give more time for limit and trip circuits to operate. However, applications that have short heat cycles, such as less than one second, require fast power turn-on. The objective is to have the power supply operate at full power for the entire heating cycle, rather than to have a large portion of the heating cycle at less than full power. The manufacturers of the contour-hardening equipment (where the heat cycles are less than 0.25 s) not only use controls designed to provide very fast power ramps, but they may even use controls that initiate turn-on at zero crossing in the line phasing. One example of fast turn-on is where RF was used to provide five heating cycles every second for automated sealing of pull-off tabs on cans.

Regulation and Control. Most induction heat treating operations are based on a predetermined level of power applied for a given length of time. Repeatability is produced through accurate timing and through a consistent amount of power produced in the workpiece every heat cycle to produce exactly the same austenitizing cycle. Electronic timers are used to provide accurate heating cycles, often through use of programmable controllers. The requirement is that each workpiece must be processed so that the same total power is induced with the same heating time. The process is kept in control through regulation of the output power and accurate timing. However, the advent of six sigma quality control requirements has required more advanced monitors for quality assurance. The most common monitor is the energy monitor.

Energy Monitors and Monitoring of Process Variables. The basic energy monitor measures and displays the total energy delivered in the induction coil in kilowatt-seconds. Voltage and current feedback signals are fed from a point (the coil is the accurate location) into a monitor that displays the kilowatt-seconds used during the heat cycle. Once a workpiece heating pattern is developed with a heating cycle, the monitor can have upper and lower kW-seconds limits entered into the control system so that there will be a system fault if the proper number of kW-seconds is not reached. Test workpieces must be run to establish the proper limit settings. Another way that the energy monitor can be used is to actually control the length of the heat cycle through kW-seconds, with a timer providing the high limit. The most precise use of the energy monitor is one in which the kW-seconds are displayed on a monitor so that the actual kW-seconds are displayed on a curve through the heating cycle. With this type of display, the high and low limits can be characterized for each application. Figure 3.8 shows an example of monitor readings for an induction cycle with a low-power heating cycle and a high-power heating cycle in kW-seconds, with limits. Figure 3.9 illustrates readings from a kW-second monitor that has the upper and lower power limits characterized. About halfway through the cycle a problem occurred and the power dipped below the lower bandwidth setting, indicating a potential process problem.

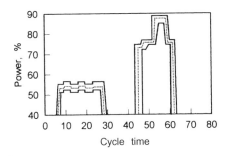

Fig. 3.8 kW-second process monitor. Source: Ref 3

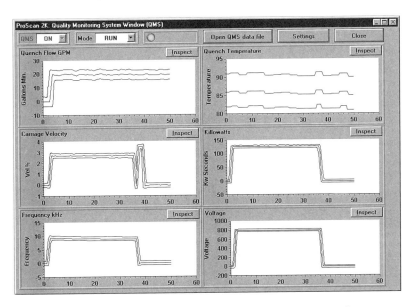

Fig. 3.9 Quality monitoring system main screen from a scan speed process monitor. Source: Ref 6

While the monitors are a useful addition to quality control systems, they may not indicate all conditions that might produce a bad workpiece. Induction coils with magnetic flux concentrators will eventually produce workpieces out of tolerance due to deteriorating or spalling of the flux concentrator. The coil will produce the same kW-seconds into the workpiece, but less energy will be induced into the workpiece while the coil losses are higher. Either periodic workpiece inspection or advanced nondestructive testing (NDT) techniques (as discussed in the Chapter "Standards and Inspection," in this book) must be used.

It is recommended that the power supply output mode monitored is not the same as the power supply control regulation mode. For instance, if a power supply is regulating on constant power, then either coil voltage or kW-seconds during the cycle should be monitored. More advanced quality-control monitors characterizing the expected output of all variables during the cycle so that high-low limits can be set for each individual process parameter. Data are collected from all variables at fixed time increments throughout the cycle and are compared with previously entered data and limits to ensure that process values always stay in limits and that the process is in control.

Other process control variables can be monitored in similar fashion, as represented in Fig. 3.10, where the variables measured and displayed include kW-seconds and coil voltage. These more sophisticated monitors are used more often with dedicated induction process systems for specific

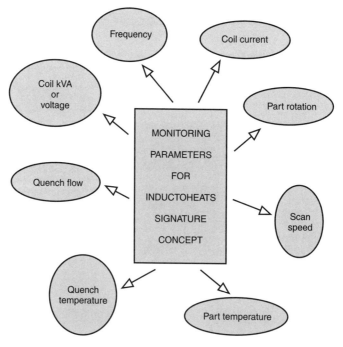

Fig. 3.10 Monitoring parameters for signature concept. Source: Ref 3

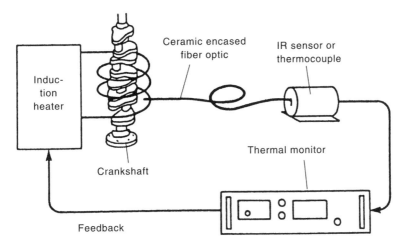

Fig. 3.11 Simple temperature control system. Source: Ref 4

parts because of the testing and qualification needed to characterize all of the variables for the process monitoring. Figure 3.10 shows measurement of other variables that can be monitored in a similar fashion. One of the variables that is sometimes measured and monitored is the workpiece temperature throughout the heating cycle.

Temperature Control through Use of Pyrometers. While experienced operators can tell by the color and scaling of the workpiece during austenitizing that the workpiece temperature is approximately correct, temperature measurement can provide precise temperature readings. Infrared pyrometers and fiber optics are most widely used for temperature measurement with induction.

Infrared pyrometers are accurate to 0.5 to 1% of the reading. They can be used as portable sensors or as part of a permanent, continuous temperature monitoring or controlling system. The pyrometer heads or fiber optic pickups need a clear line of sight onto the area at which the temperature is being measured. Where the heat cycle has fast temperature changes, a digital readout is preferred because the analog type meters have a delay in indicating the metered output.

There are three types of infrared pyrometers:

- Single-color or single-band pyrometers measure infrared radiation from a fixed wavelength. These are the most versatile because of their suitability to many applications. Their accuracy is affected by the presence of scale on the workpiece and any smoke coming off the workpiece, and by any changes in emissivity of the workpiece.
- Broad-band pyrometers measure the total infrared radiation emitted by the target. Similar to single-color pyrometers, the accuracy of these pyrometers is also adversely affected by the presence of scale and smoke.

- Two-band or two-color pyrometers measure the radiation emitted at two fixed and closely spaced wavelengths. The ratio of the two measurements is used to determine the temperature of the target. The adverse effects of other two pyrometers are largely eliminated by this pyrometer. Because of its accuracy this is the most widely used pyrometer. It is readily adaptable for measuring temperature in induction heating systems using a variety of optical systems for focusing, including fiber optics.

Figure 3.11 shows a system schematic for the use of an infrared pyrometer during the induction hardening of a crankshaft. Systems such as this may be either closed loop, with the pyrometer controlling the heating cycle, or open loop, with the pyrometer acting as an over-temperature shut-off control. Most pyrometers are used for over-temperature control.

Diagnostic systems help to show the operating condition of the induction heating system. Table 3.2 shows the typical capabilities of diagnostic control. With planning, the status of virtually every component of an induction system can be displayed with feedback for diagnostics where

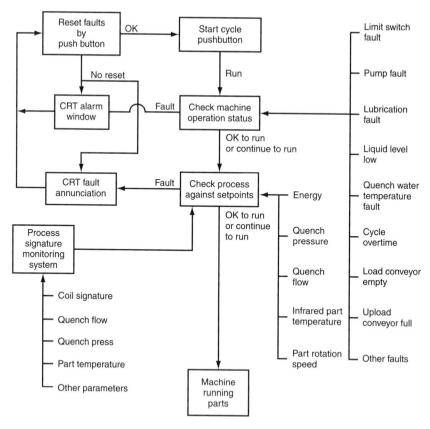

Fig. 3.12 Flow diagram of a diagnostics control plan. CRT, cathode ray tube. Source: Ref 7

necessary. Figure 3.12 shows a flow diagram of a diagnostics control plan. The more advanced control systems allow the entry of all process parameters to be characterized in real time on a screen with limits set. Operators can enter, verify, and monitor power levels, heating time, the power supply output during the cycle, part temperature during the cycle, scan speeds during the cycle, and other conditions as desired, such as quenchant pressure, quenchant flow, and part rotation speed.

Window of Full Power Output. Figure 3.13 shows the effect of tuning to produce a frequency that results in full power. Once the resonant frequency is reached, the power output drops rapidly. (This is known as "going over the hill.") The width of the frequency range that will produce rated power represents the "window" within which the power supply can be tuned to make rate power. The ability of a given power supply to make full power into a given work load and output impedance depends on the system design and application. However, the ability of different types of power supplies to broad band tune and to make full power into different loads and coil impedance varies widely. Some power supplies need to have 100% over rating of power to essentially be a broad band-tuning unit. The ability of given power supplies to broad band tune can sometimes be increased by installing additional tuning capacitors, wider-range output transformers, and the use of auto transformers in the heat stations.

Table 3.2 Typical capabilities of diagnostic equipment

Machine diagnostics	Process diagnostics	Process signature capabilities
• Proximity switch failure	• Power level	• Quench pressure
• Limit switch failure	• Heat time	• Quench flow
• Liquid levels	• Scan speed	• Power level
• Liquid temperatures	• Rotation speed	• Voltage level
• Inductor ground	• Part temperature	• Current level
• Power supply status	• Energy into a part	• Rotation speed
• Pump operation status	• Quench flow	• Part temperature
• Guard/safety gate status	• Quench pressure	
• Coil water cooling pressure	• Quench temperature	
• Coil water cooling flow	• Quench concentration	
• Part position gaging		
• Part present indication		

Source: Ref 7

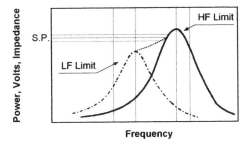

Fig. 3.13 Power supply output at resonant frequency. SP, set point; LF, low frequency; HF, high frequency. Source: Ref 3

36 / Practical Induction Heat Treating

If in doubt, the recommendation is to purchase a power supply with reserve power. It is easier to produce 100 kW at 50% with a 200 kW power supply than it is to run the 100 kW power supply at full power over all conditions.

Other conditions that should be reviewed in new equipment are overall equipment reliability, maintainability, and the service and parts organization.

Fixtures and Workhandling. Simple fixtures, such as worktables, sit in front of the heat stations and may require no interconnection. On the other hand, the fixtures and workhandling equipment may be very elaborate. Descriptions of lift-and-rotate mechanisms (example shown in Fig. 3.14), index tables, and scanners are described in the Chapter "Induction Heat Treating Process Analysis" in this book. These devices provide whatever is necessary to locate the part in the induction coil and the means for moving parts into and out of the coil. There are many standard types of designations for this equipment, and there have been many unique mechanisms.

The following is a list of design factors that should be considered in the design of fixturing:

Fig. 3.14 Picture of lift-and-rotate mechanism. Source: Ref 8

- The workpieces must be safely held in fixtures or nests that do not interfere with any part of the heat treating process. For example, nests that hold the parts may need to be made from nonmagnetic materials, or even cooled during and after hardening.
- Workpieces must be positioned positively with the degree of accuracy needed for each step of the process. Mechanized systems may need proximity devices to make certain that the workpiece is accurately located in the coil before the workpiece is positioned into the coil.
- Workpieces must be positioned properly and supported well enough to minimize distortion. For example, small-diameter workpieces held between centers can distort with excessive center pressure. Long workpieces may need mechanical restraint on each side of the coil.
- Frequencies below 10 kHz can cause some workpieces to levitate. The fixturing design must prevent levitation.
- Electrical conducting materials used in construction of the fixture should be kept away from the heating field; and stray, induced currents should be avoided in drives, bearings, and moving parts.
- Bearings and moving parts that are under the quenchant must be sealed to prevent entry of the quenchant into these bearings and parts.
- The necessary operating controls for cycle start and emergency stop must be in reach of the operator. Also on induction hardening systems that have frequent setup change, it is desirable to have the power supply meters in sight of the operating position. This helps to facilitate safe tuning during set up.
- The working height of the fixture should be at a level conducive to efficient flow of workpieces in and out of induction hardening.
- Operational safely concerns as required by OSHA regulations must be met to prevent harm to the induction heater operator.

Quench Systems

Quenchant selection is discussed in the Chapter "Quenching" in this book. Quench systems can vary from the simple use of line water, with the spent water going down the drain after the workpiece is quenched, to more sophisticated quenching systems with tanks, heat exchangers, pumps, filters, and temperature regulation controls.

In the past, dunk oil quenching and water-soluble oils were widely used. Today, although oil is still used for some applications, polymer quenches have widely replaced most of the oil quenches. Polymers are not flammable, so they provide no fire hazard. In addition, polymers are easily cleaned off after quenching. Polymer quenchants contain biocide and rust inhibitors. The vapor pocket reduction provided by low polymer-to-water ratios has been found to make polymer quenchant more desirable than straight water quenchants in many applications.

Proper quench system design requires consideration of many factors during the selection of a quenchant quench system:

- The quenchant must be selected for the characteristics desired. This includes the quenchant speed, desired operating temperature range, and additives. Some biocide additives contain chlorine, which can combine with alkaline glycol to form cholorethel. If nitrosamine is prohibited, end users can also specify nitrate or non-nitrate based polymer quenchants to avoid the formation of nitrosamine. Because the smell of the quenchant can be obnoxious, venting of the quench tank is needed in cases where there is considerable vaporization.
- The quench system must be sized for both the volume of quenchant and the heat extraction needed. A source of line water, with given flow and input temperature, is needed for cooling the quenchant in the quench system heat exchanger. Sizing of the system is a function of the line water and quenchant flow, the temperature differentials of the line cooling water and the quenchant, and the size and heat removal capacity of the heat exchanger. Cooling towers, chillers, and city or well water systems can be used for the line water. (Chillers are expensive to operate.) Due to cost, in most locations cooling towers are used when substantial line cooling water is needed. Cooling towers have the limitation of not being able to cool the recirculated water below high outside temperatures and high humidity. As with the quenchant system, closed systems involving chillers or cooling towers may need to have biocides added, and they may need filtration to remove debris.
- The quench system must be able to provide the amount of quenchant needed at the coil. Design of quenchant outlets in coils, quench rings, and quench pads will be discussed in the appendix "Quench System Design" in this book. The system design must be capable of maintaining the quenchant within the specified control range, as the range should normally not exceed more than 7 °C (10 °F). A typical control range is from 35 to 38 °C (90 to 100 °F). If the control range is higher than room temperature, heaters may be needed to bring the tank up to temperature. An over-temperature alarm or cutoff to prevent high-temperature operation is always recommended.
- The quench tank and piping must not react with the quenchant. Polymers are very hard on painted tanks. In addition, and to prevent galvanic corrosion, the piping should not be made of dissimilar materials. A tank capacity of 3 to 4 times the quench flow per minute is recommended.
- The quench system piping, solenoids, and manifold must be designed so that full volume of quenchant is supplied to the quenching position on command. Care must be taken so that the quenchant itself does not drain out of the quench ring. Ideally, when a quench solenoid opens,

full flow must occur with no necessity to fill the quench ring itself before the quenchant can emerge.
- Suitable filtration must be put in the quench tank. The quenchant must be kept free of foreign material and scale. Y-Type Strainers are widely used because of the low cost. These are basket type strainers, and they serve to trap only the larger particles. Cartridge and bag type filtration units can be specified to filter from 5 up to 100 μm. A differential pressure switch across the inlet and outlet sides will detect a clogged condition. If the filters are placed in a bypass configuration (similar to the oil filter in a car), separate flow is run through the cartridge continuously. In this case a flow switch can be used to indicate the flow. A wide variety of controls and systems are available to help "foolproof" these systems. When quench flow is being monitored as one of the process variables, the usual requirement is for monitoring the pressure at the pump outlet or manifold, and for monitoring the quenchant flow into the quench ring.

Continuous automatic, indexing media-style filters work well for users who experience problems with manual replacement of filter media. These are sometimes called dragout style filters because the filter paper is dragged across a perforated plate. When the pressure differential is too high, the filter is automatically indexed into a hopper. These filters are good for applications that are high in scale and chips because their large volume and automatic indexing assures good filtration. Positive pressure is used for filtration, and a "clean tank" is required on applications that cannot afford to have the quenchant flow disrupted for cleaning the filter.

Load Matching

It is necessary to match or tune the load impedance with the output impedance of the power supply in order to obtain the delivery of full power from the power supply to the workpiece. Electrically, impedance is the vector sum of the resistance, current, and voltage. Coil voltage is normally used for matching reference. Figure 3.15 shows illustrations of voltage matching, starting with matching a 6 V light bulb to a 120 V household circuit to an 80-volt output from an induction power supply. If the induction coil shown has a 30 V requirement and the power supply has an 800 V output, then a transformer can be used to match the coil to the power supply. The electrical components used for load matching, such as transformers and capacitors, are located in the heat station. Heat stations may be either built into the power supply or furnished separately. This book will not discuss older technology power supplies used for heat treating such as motor generator sets and spark gap units. More information for these power supplies can be obtained in Ref 9 and 4.

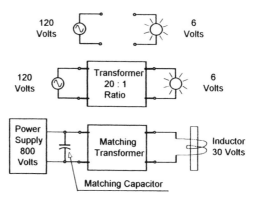

Fig. 3.15 Load matching-impedance matching. Source: Ref 3

When purchasing equipment or when designing for new applications, it is important to know the capability of the particular power supply in order to match to the proposed induction coil. Because of the significant differences in power supplies, the specifications of the individual power supply must be reviewed. There is a wide variety in the capabilities of different power supplies, as furnished by different manufacturers, to produce full power over a wide range of different loads. While this may not be significant in the case of the power supply and process that are designed for one specific part, it may be significant where different types of parts need to be induction hardened and versatility is needed. The ability to produce full output is dictated by the type of power supply and heat station, the transmission line to the induction coil, and the induction coil and load. Power losses occur in the transmission of power from the terminals of the heat station to the induction coil. Coil lead lengths, lead design, and transmission line design must be considered in equipment layout.

Tuning of Solid-State Inverters

A discussion of power supplies is not complete without mention of the most dreaded work to many—tuning. However, tuning is not hard to do if the proper sequence is performed. The tuning of solid-state is easier than the motor generator sets used before solid-state because of the faster limit and control circuits in solid-state. Before tuning, set up a log for recording the meter results during tuning, so that any changes in meter readings can be used to decide what to change for the next step. Optimal tuning requires a combination of using transformer taps to tune for voltage while using capacitor taps to tune for a frequency range that will permit the voltage match. The tuning on the floor involves trial and error, although in the future it is expected that methods will be available for using logic in the control circuits to help with tuning.

The object of tuning is to match the voltage and current to achieve maximum power at the desired frequency. The meters are used to monitor the output for tuning purposes. Ideally the instruments should indicate power, voltage, current, and frequency. None of these should be permitted to operate in the limit condition because the power supply will not be capable of producing an output into the coil when a limit condition is occurring. On solid-state power supplies that do not have a current meter, keep in mind that, assuming power factor is a constant, the output power is equal to voltage time current (kW = VA) so that the current output can be estimated for tuning purposes. To tune:

1. Use the tuning for a similar part if available. Start with a low amount of capacitance tapped in and the power control set at a low level. Turn on the heat and observe the meters. If none of the meters is running into a limit, slowly increase the power until an observation can be made at which the meter reads higher than the others (i.e., is the voltage running high while the current is low?). Turn the power off and record the readings. If the power is sufficient to overheat the workpiece, cool it before reheating. Care must be taken not to overheat the workpiece during these tests.
2. If the voltage is high, take out some turns on the transformer ratio. For example, go from a ratio of 16 to 1 to a ratio of 12 to 1. If the current is high, add turns to the transformer ratio. If the frequency is high, add capacitance. On power supplies with a large frequency range, match voltage and current through the transformer before attempting to force the frequency change through the addition or subtraction of capacitance.
3. After changing taps, turn the power down again and repeat the low power step, gradually increasing the power until either full power or a limit is reached. Again, record the readings.

Repeat these processes until the voltage and current readings are similar and the frequency is within the desired range.

Theoretical tuning can be done by using a load frequency analyzer, as applied to a given transformer, to work coil match that determines the resonant frequency of the induction heater without heating the workpiece. In this case, the transformer matching ratio is either calculated or estimated by rule of thumb. After obtaining the resonant frequency, capacitance is added or subtracted from the circuit to match the tuned frequency of the work station to the rated frequency of the power source.

Radio frequency, oscillator tube power supplies may use tap changes in the tank coil or a means of adjustment of the grid current output for tuning purposes. The power supplies require that the grid current be kept between specified range at full power. These power supplies tend to have wider bands of impedance matching than solid-state power supplies. If tuning is needed during setup, it is generally very easy to do. Stepless power control is usually accomplished through the use of silicon controlled rectifier power controllers that vary the output power through variation of the input line voltage. However, depending on the particular circuit used, the operator can use changes in grid current or taps on the tank coil circuit to change the output power.

REFERENCES

1. R.E. Haimbaugh, Induction Heat Treating Corp., personal research
2. D.L. Loveless et. al., Power Supplies for Induction Heating, *Ind. Heat.*, June 1995
3. V. Rudnev et al., *Steel Heat Treatment Handbook*, Marcel Dekker, Inc., 1997
4. S.L. Semiatin and D.E. Stutz, *Induction Heat Treating of Steel*, American Society for Metals, 1986
5. T. Boussie, "Water System Problems and Solutions," paper presented at 6th International Induction Heating Seminar (Nashville, TN), Sept 1995
6. Induction Systems, unpublished data
7. R. Meyers, Diagnostics, a Powerful Tool for Induction Heating, *Adv. Mater. Process*, July 1992, p 41
8. Ajax Magnathermic, unpublished data
9. C.A. Tudbury, *Basics of Induction Heating*, John F. Rider, Inc., New Rochelle, NY, 1960

CHAPTER **4**

Induction Coils

THE COIL, also known as inductor or induction work coil, is basically a transformer primary that induces high-frequency output of an induction power supply into a workpiece, which is effectively the transformer secondary. Coil design is application specific, so the type of coil to be used should be selected before designing fixturing. The type of process used, such as whether the workpieces are heated single shot or scanned, influences coil selection. Coil design and construction principals are discussed in the Appendix "Scan Hardening" in this book. This Chapter will deal with general theory, types, and applications of coils.

Classifications of Coils by Electromagnetic Field

The two classifications relate to the direction of eddy currents produced by the coil in the workpiece:

Longitudinal flux (reverse current flow) induction coils are by far the most widely used type of coil, with solenoid types of coils most commonly used. Longitudinal flux coils should not be confused with channel coils, which are orientated in the longitudinal direction of a workpiece. The workpiece is surrounded or enveloped, with the turns on opposite sides so that induced current flows around the workpiece as shown in Fig. 4.1. When the air gap between the coil and workpiece is reasonable for the frequency and load conditions involved, heating can be quite efficient because the flux lines tend to be confined. Figure 4.2 illustrates the confinement of the flux field between two conductors with opposing current. Solenoid coils come in many different forms, shapes, and adaptations, with the circular or enveloping coil the simplest form. As previously discussed in the theory of induction in the Chapter "Theory of Heating by Induction" in this book, solenoid types of coils require the workpiece thickness to be at least four times the reference depth for efficient operation. The coils can be oriented so that long workpieces are either encircled by a solenoid coil with the current flowing around the circumference of the part, or, in the case of channel coils, oriented perpendicular to the workpiece so that the current flows along the longitudinal direction. Solenoid coils have many subclassifications, which are discussed in this Chapter.

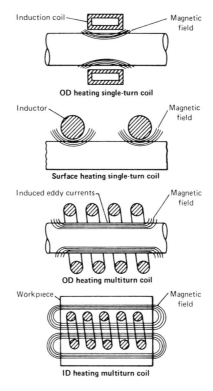

Fig. 4.1 Magnetic fields and induced current produced. Source: Ref 1

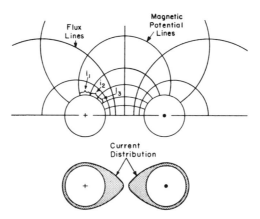

Fig. 4.2 Flux plot of two parallel conductors carrying current in opposite directions. Ref 2

Transverse flux (sometimes called proximity) coils are not as widely used and are used to heat work pieces where the cross-section thickness is less than four times the reference depth. Figure 4.3 shows an example of a transverse coil. The workpiece is essentially placed in between turns of the induction coil, in which the current is flowing in the same direction. By doing this, there is no current cancellation effect and thin pieces can be heated efficiently.

Coil Design

Factors that influence coil design and selection include the dimensions and shape of the workpiece, number of workpieces to be heated, hardness pattern desired, production process to be used (such as single shot or scanning), frequency and power input, how the workpiece is to be quenched, and coil-life considerations. Table 4.1 shows the efficiency of different style coils that do not have flux concentrators applied. In some circumstances, depending on the process used, different types of coils can be used. The same part may be scanned on a lower power system that is single-shot hardened on a high power system. Many coils are built based on knowledge of coil design principles and on experience, while other coils

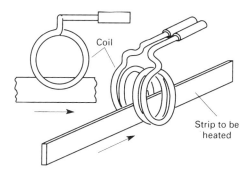

Fig. 4.3 Example of transverse flux coil or figure showing current. Source: Ref 3

Table 4.1 Efficiency of different types of coils

Type of coil	Coupling efficiency at frequency of:			
	10 Hz		450 kHz	
	Magnetic steel	Other metals	Magnetic steel	Other metals
Helical around workpiece	0.75	0.50	0.80	0.60
Pancake	0.35	0.25	0.50	0.30
Hairpin	0.45	0.30	0.60	0.40
One turn around workpiece	0.60	0.40	0.70	0.50
Channel	0.65	0.45	0.70	0.50
Internal	0.40	0.20	0.50	0.25

Source: Ref 4

need development programs in order to produce conforming workpieces. Computer simulation for coil modeling is discussed in the Appendix "Scan Hardening" in this book.

Coil design considerations for efficiency and use include:

- The coil should be coupled as close to the workpiece as possible. Table 4.2 shows recommended coil-coupling distances. The coil-coupling requirements may depend on other factors, such as how the part is to be loaded or positioned in the coil, and the shape of the area to be heated.
- The leads into the coils must be designed for the best efficiency.
- The coils must be designed so that they are rigid and do not move when power is applied.
- The coils must be designed so that they do not overheat during use and do not develop stress fractures during use. As a rule, lower frequency coils generally operate with higher power inputs, so they require more rigidity and better cooling.
- All coils have significant power losses and need good cooling. Separate high-pressure cooling systems may be necessary for high-power applications.
- The coil must be designed so that the magnetic flux lines produce eddy currents that heat the desired areas. The highest concentration of flux lines is inside the coil, producing the maximum heating rate there. Coils can be made with different contours and shapes or can be made with more than one turn. Radio frequency (RF) coils need closer coupling and concentricity around workpieces than the lower frequencies. Flux concentrators are used to increase coil efficiency or to concentrate the flux to specific areas in the workpiece.

Magnetic flux concentrators or intensifiers are used very effectively to increase the efficiency of some coils. In other cases, flux concentrators are necessary to produce the pattern required in the workpiece. Figure 4.4(a) shows the current distribution in a conductor. When a workpiece is placed under the conductor, the current is redistributed as shown in Fig. 4.4(b). Because of proximity effect, a significant part of the conductor's current flows near the surface of the conductor that faces the workpiece. The remainder of the current is concentrated on the sides of the conductor

Table 4.2 Recommended coil coupling distances

Frequency (kHz)	Coupling to workpiece	
	in.	mm
1–3	0.12–0.24	3–6
10–25	0.08–0.15	2–3.8
50–450	0.06–0.09	1.5–2.25

Source: Ref 5

with a current induced in a general area in the workpiece close to the conductor. After a flux concentrator is placed around this conductor, as shown in Figure 4.4(c), practically all of the conductor's current is concentrated on the surface facing the workpiece. A significantly larger current is induced in the workpiece directly next to the conductor. Some applications such as bore heating show substantial increase in efficiency. Correctly applied, the benefits are:

- Reduction of operating power levels required to obtain the desired heating of workpieces
- Improvement of process efficiency and decrease in the amount of energy used

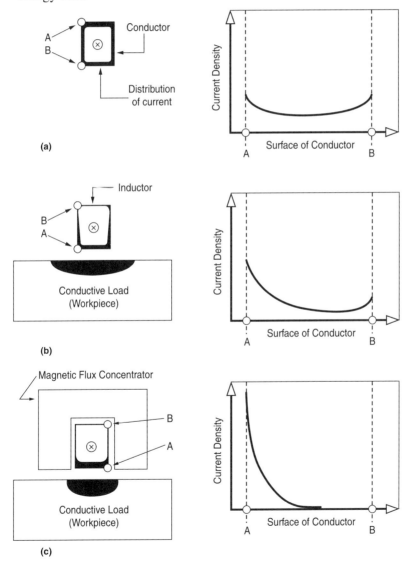

Fig. 4.4 Current distribution in straight conductor. Source: Ref 6

- Increasing the productivity rate of the heat-treating system
- Heat pattern improvement because of the control of the distribution of heat in the workpiece
- Protection of the workpiece or fixture components against unintended heating by stray flux
- Improvement in coil load matching to the power supply

The use of flux concentrators usually provides several of the previously mentioned benefits. Table 4.3 shows the efficiency of internal coils with and without concentrators. When the power supply is tuned for the same coil voltage, the bare coil has a coil efficiency of 68%, as compared to an efficiency of 87% of the coil with flux concentrators. Figure 4.5 shows increases in case depth that can be with a 28 mm (1.13 in.) ID by 6.25 mm (0.25 in.) wide coil heating a 1 in. round bar with a 380 kHz and a 8.3 kWh power supply at the energy levels shown. At the same energy level the coils with concentrators had increased case depths over the coils without concentrators. The technical and economical significance of each effect depends on the specific conditions of the particular application. Using computer simulation and full-scale experimentation, the Centre for Induction Technology has conducted detailed studies on the influence of flux concentrators (Ref 9). The results show that the

Table 4.3 Effect of flux intensifiers on ID coil efficiency

	Coil voltage, V	Workpiece power, kW	Coil losses, kW	Coil efficiency, %	Power factor, kVA	Apparent power
Power Coil	88.6	68.0	10.4	87	0.22	354
Bare Coil	30.6	8.4	4.0	68	0.10	122
Bare Coil(a)	87.0	68.0	32.4	68	0.10	992

Note: (a) Bare coil with the same workpiece power as the Power Coil. Source: Ref 7

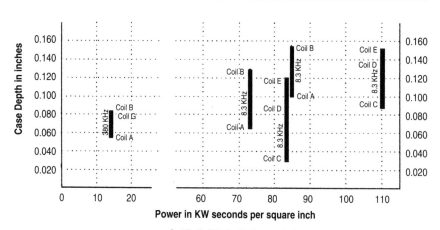

Fig. 4.5 Increase in case depth with concentrators in inside diameter (ID) coils. Source: Ref 8

proper application of flux concentrators is always beneficial in an induction heating process.

In many cases the requirements needed for magnetic materials used for flux concentrators for coils are very severe. The materials must work in a very wide range of frequencies and must possess high magnetic permeability and saturation flux densities. When magnetic materials saturate, they lose their ability to concentrate magnetic flux. Stable mechanical properties with resistance to the elevated temperatures caused by hysteresis and heat transfer from the workpiece are important. Because of the nature of heat treating processes, the material must withstand an attack of hot water, quenchants, or active products. For successful application of flux intensifiers machinability is also a very important property because it enables the intensifiers to be contoured to the shape of the coil. Three groups of magnetic materials are used for intensifiers: laminations, ferrite, and magnetodielectic materials.

Laminations are manufactured from grain-oriented nickel-steel and cold-rolled and hot-rolled silicon-steel alloys. Mineral and organic coatings provide the insulation of the laminations. The lamination thickness used for induction coils ranges from 0.1 to 0.6 mm (0.004 to 0.024 in.), with the thin laminates being used for the highest frequencies. The main frequency use of laminations application is below 10 kHz, although at times they are used at frequencies up to 30 kHz with reduced flux density and intensive cooling. Laminations have the highest possible permeability and saturation flux density, so at frequencies below 10 kHz they are the most efficient. One of the main advantages of laminates is that they can withstand higher temperatures than other materials. Laminations need to be tightly stacked together. They have overheating problems when they are stacked in sharp radii such as the ends of single-shot coils.

Ferrites have high permeability only in weak flux fields with low losses. Ferrites work in a wide range of frequencies when properly selected and applied. The drawbacks that restrict application as a good material for flux controllers include low saturation flux density and Curie temperature, high sensitivity to thermal and mechanical shock, and the inability to be machined except by grinding and cutting with diamond tools.

Magnetodielectric (electrolytic and carbonyl iron-based) materials are made from soft magnetic particles and dielectric materials that serve as a binder and electric insulator of the particles. Different grades of the materials are currently used for most applications. The materials are produced by compacting different magnetic powders and binders, followed by thermal treatment. The compacted materials are machinable with standard tools, and they have excellent thermal, chemical, magnetic, and mechanical properties.

Moldable materials made from insulated micro iron particles suspended in a polymer base have been developed. These materials are sold in a soft, flexible form. A catalyst is added before forming on a coil, and the coil and composite are baked in a low-temperature oven.

The disadvantage of flux intensifiers is that substantial cost can be added to the coil. In addition, coil life is reduced. Bare coils without concentrators have long life expectancies and usually fail because of a workpiece hitting the coil or an arc. Coils with concentrators have shorter life cycles, with some producing as few as 10,000 parts before failure. When coils with concentrators fail, the failures may be very gradual over a period of time, or they can be catastrophic. Dedicated production lines using coils with concentrators often have a number of coils on hand so that coils can be removed and rebuilt when necessary. Many induction processes use flux concentrators only when the overall benefits exceed the disadvantages. Bare coils work quite well in many applications.

Coil Characteristics Versus Frequency

Coils in the lower frequencies of 1 to 10 kHz can have mechanical vibration that requires rigid restraint and mounting techniques, particularly in the 1 kHz region. Close space multi-turn coils will spread apart when power is applied unless they are restrained. These lower frequencies are used for deeper case hardening and for larger diameter through heating applications. The coils used for heat treating in the 3 to 25 kHz frequency range tend to be low-turn coils, although coils used for continuous heating of bars and tubes can use multi-turn coils such as those used in the forging industry. Machined coils are more likely to be used with the low frequencies, while coils made from tubing are more likely to be used with high frequencies. Figure 4.6(a) shows the current produced in three areas in relationship to coil position at the end of a bar for high frequency, while Fig. 4.6(b) shows the same effect for low frequency. High frequency has higher edge effect, which is an effect in which there is more current concentration at the end of a bar that produces deeper heating. This can affect coil selection from the viewpoint that the higher frequencies are more limited in the use of wide or multi-turn coils in scanning operations due to this edge effect tendency.

Coil characterization is the technique of adjusting the coil design so that the induced eddy currents produce a uniform heating pattern in the workpiece. The number of turns, the spacing between turns and the workpiece, the power density, and the frequency determine the heating effect and final austenitized pattern. Figure 4.7 shows the effect of coil placement at various locations, both on the outside diameter (OD) and inside

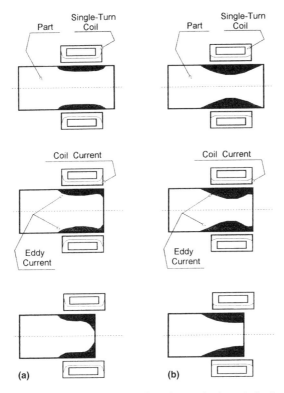

Fig. 4.6 Electromagnetic end effect for (a) high and (b) low frequency.
Source: Ref 10

diameter (ID) of round parts. The position and overlap of the coil affects the heating effect at the edges. Coils are designed with overlap and then positioned so that the heating at the edge produces the same depth of pattern as produced in the center. The lower the frequency, the more overlap is required because of the edge-heating effect. Standard close-coupled coils with flux concentrators operating at 10 kHz require approximately 3 mm (0.120 in.) overlap, while only 1.5 mm (0.060 in.) overlap is required for 450 kHz. The closer the coupling between the turns and the workpiece, the more intense the magnetic flux field. Intense flux fields produce more intense eddy currents and, consequently, faster heating. Machined coils can have different machined ID dimensions in the same coil that affect the energy input into the workpiece, similar to the effect that can be accomplished by winding multiturn coils with the turns having different IDs. Table 4.3 shows the efficiency of different types of induction coils. Note that the efficiency increases when two turns are used to heat the same width instead of one turn.

Coil Leads and Dead Spots. The magnetic flux field in a coil is not entirely symmetrical because the points at which the coil leads are attached distort the flux field. The effect is that the flux field is off center. This effect is more significant with the RFs and with single-turn coils.

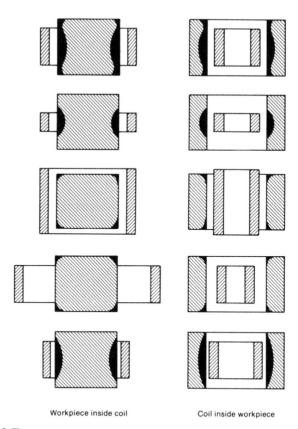

Fig. 4.7 Effect of relative proportions of workpiece and coil. Source: Ref 11

Rapid rotation of the parts helps to produce uniform heating of the workpiece. In short cycles, higher speed rotations of up to ten times in the cycle are necessary, while for time cycles of 4s or longer, a minimum rotation speed of at least 2 rps is recommended.

Coils for Static Heating. Static heating is heating the workpiece in place with a timed heat cycle. The coil shape depends on the geometry of the workpiece or part being heated. Round parts such as fine-pitch gears use encircling coils. Other parts, such as the tips of tools or specific-wear areas, use odd-shaped coil designs, such as those shown in Fig. 4.8, which are individualized to the type of area to be heated.

Coils for Single-Shot Heating. Single-shot and static heating are often used to define the same type of heat cycle. Coils for high intensity, single-shot heating are designed to locate horizontally on the workpiece so that current is induced in the longitudinal direction. Round workpieces are rotated at high speed so that a case can be produced around the diameter. Coils for single-shot heating generally require characterization and usually have flux concentrators to help improve the coil efficiency. The size limitation of a workpiece for single-shot hardening is limited by the

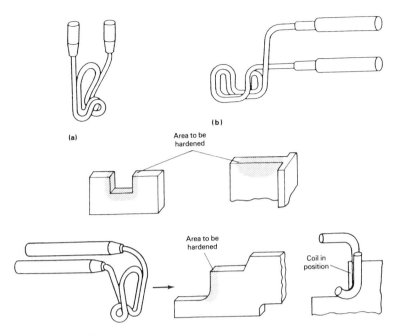

Fig. 4.8 Localized heating coils. Source: Ref 3

amount of power available in the power supply. Single shot is used on large workpieces when high production rates justify the equipment cost. Coils for single-shot hardening have coil designs that permit either quenching in place through the coil or movement of the part out of the coil for quenching.

Coils for Scanning or Progressive Hardening. Scanning requires the workpiece to delay in position at the start of the heating cycle, then to have the workpiece start moving through the coil, progressively heating, and quenching. The major difference in coil design between static heating and scanning occurs when quenching is integrated into the coil with separate chambers built into the coil for the quenchant and for the cooling water for the coil. Depending on the application, both encircling and longitudinal coils are used. The most common scanners are fixtures designed for vertical scanning. Figure 4.9 shows two coils connected in series heating workpieces on a dual spindle, vertical scanner.

Basic Types of Coils

Solenoid encircling coils are the most efficient coils used with typical applications such as heating the outsides of workpieces. The number of turns depends on the load-matching characteristics of the power supply (as discussed in the Chapter "Induction Heat Treating Systems" in this book), the heat treating pattern required, and the process requirements.

Workpieces with high production requirements tend to use much larger coils that require more power to produce the required power density. Multi-turn coils tend to be used more with the RFs, while one- and two-turn coils tend to be used with the lower frequencies. Solenoid coils are designed with and without integral quenches and are also used for scanning. Examples of these coils are shown in the following figures:

- *Fig. 4.10:* Single-turn coil with rigid construction, as commonly used for medium and low frequency, that was designed for annealing a welded tube. The same coil design could be used for heating any round shape, gear, or sprocket of appropriate width. Quenching is done externally to the coil.
- *Fig. 4.11:* Single-turn coil with rigid construction and integral quench. This coil would be used for static heating, with quenching in the coil following austenitizing.

Fig. 4.9 Dual spindle scanner with coils. Source: Ref 2

Fig. 4.10 Single-turn inductor. Source: Ref 12

Fig. 4.11 Single-turn outside diameter (OD) single-shot inductor. Source: Ref 12

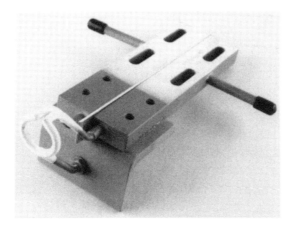

Fig. 4.12 Single-turn radio frequency (RF) tubing coil. Source: Ref 12

Fig. 4.13 Single-turn machined coil. Source: Ref 12

- *Fig. 4.12:* Single-turn RF coil with stud providing rigidity and aluminum oxide protective coating. This particular coil was designed for induction brazing. The same coil could be used for a number of different types of applications, such as austenitizing the end face of a workpiece.
- *Fig. 4.13:* Single-turn machined coil with integral quench designed for rigidity. The ID of a workpiece is austenitized followed by jogging into position from the integral quench.
- *Fig. 4.14:* Single-turn machined coil with integral quench for scanning designed for close impingement of quenchant after scanning. The machined taper at the quench holes provides the desired angle of quenchant impingement. This particular coil was designed for axle shaft hardening.

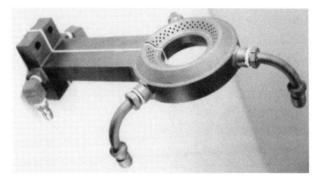

Fig. 4.14 Single-turn scanning coil. Source: Ref 12

Fig. 4.15 Single-turn scanning coil with quench. Source: Ref 12

- *Fig. 4.15:* Single-turn machined coil with integral quench for scanning designed with quench holes in the bottom. This coil was designed for splined shaft hardening. The method of attaching the outer plate at the quenchant outlet permits easy access for cleaning the quenching orifices.
- *Fig. 4.16:* Single-turn tube coil shaped for single-shot, contour hardening. This coil was designed for thin wall tube hardening and tempering.
- *Fig. 4.17:* Single-turn machined coil with integral quench for static heating and hardening gear teeth on a shaft
- *Fig. 4.18:* Single-turn machined coil contoured to shape of workpiece for scan hardening a gear rack
- *Fig. 4.19:* Single-turn machined coil with integral quench contoured to shape for static hardening a cam
- *Fig. 4.20:* Two-turn tubing coil such as commonly used for RF scan heating of a tube
- *Fig. 4.21:* Two-turn series coil with quench for scanning to produce deep case hardening of a shaft

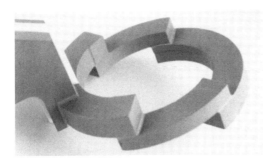

Fig. 4.16 Single-turn, single-shot-contoured coil. Source: Ref 12

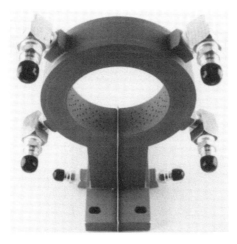

Fig. 4.17 Single-turn machined static heating coil. Source: Ref 12

58 / Practical Induction Heat Treating

Fig. 4.18 Single-turn machined contour to part coil. Source: Ref 12

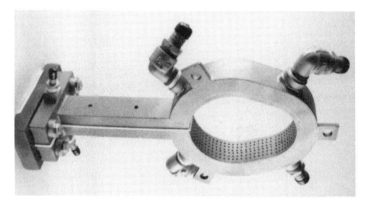

Fig. 4.19 Single-turn machined contour coil, integral quench. Source: Ref 12

Fig. 4.20 2-turn tubing coil. Source: Ref 12

- *Fig. 4.22:* Low-turn shaped coil from tubing with rigid mounting for tempering a wheel spindle
- *Fig. 4.23:* Multi-turn shaped coil from tubing coated with plastisol for dielectric compound for RF heating of nuclear fuel
- *Fig. 4.24:* Multi-turn coil with rectangular tubing for through heating prior to forging. The same coil can be used for through heating for hardening with quenching external to the coil.
- *Fig. 4.25:* Multi-turn coil for RF; direct tank connection (high voltage) for preheating above a fluidized bed
- *Fig. 4.26:* Multi-turn rectangular tube coil with flux concentrators for hardening a brake pin

Fig. 4.21 Two-turn series coil with quench. Source: Ref 12

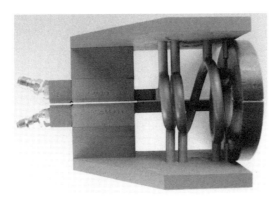

Fig. 4.22 Multi-turn tube-shaped coil with concentrator. Source: Ref 12

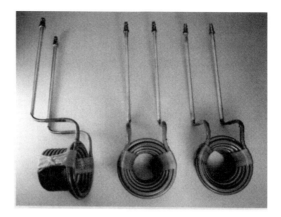

Fig. 4.23 Multi-turn shaped coil. Source: Ref 12

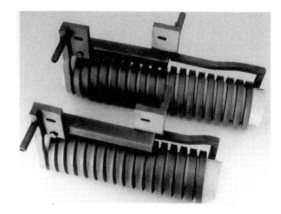

Fig. 4.24 Multi-turn solenoid coil. Source: Ref 12

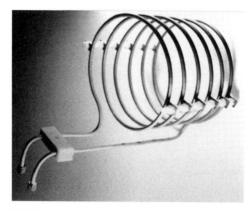

Fig. 4.25 Radio frequency (RF) multi-turn coil. Source: Ref 12

Channel or Hairpin Coil. These coils are in effect solenoid coils, flattened and elongated. Most channel coils are single-turn coils. Workpieces can be oriented both perpendicular and longitudinal to the coil and with proper fixturing a number of workpieces at one time. When the ends are cupped, this type of coil can be used to heat parts that are trayed or conveyorized. When a workpiece is inserted longitudinally, round parts are rotated and heated single shot. Workpieces that have edges, such as cutters, can be heated by this coil design. Examples of these coils are shown in the following figures:

- *Fig. 4.27:* Single-turn channel, curved coil with flux concentrator so parts on a wheel can pass through. This coil was used for the continuous heat of bolt threads.
- *Fig. 4.28:* Multi-turn channel coil designed for high-voltage terminal connection designed for tempering axle shafts.

Fig. 4.26 Multi-turn coil with concentrators. Source: Ref 12

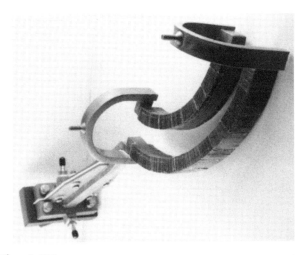

Fig. 4.27 Curved channel coil with concentrators. Source: Ref 12

Internal Coil. Single- or multi-turn coils can be used to heat the bore of a workpiece. Internal coils tend to be less efficient, particularly at small diameters at which the current path in the coil must come up through the lead in the center of the coil. The operating efficiency of internal coils tends to be the lowest, but flux concentrators help greatly to increase the operating efficiency. It is difficult to wind a coil under 12.5 mm (1/2 in.) in overall diameter; however, special hairpin coils have been designed to fit into smaller bores. For all practical purposes 12.5 mm (1/2 in.) is the smallest bore that can be heated with 450 kHz, while 250 mm (1 in.) is the smallest bore that can be heated with 10 kHz. Examples of these coils are shown in the following figures:

- *Fig. 4.29:* Single-turn machined coil with integral quench designed for ID scanning of cylinder liners

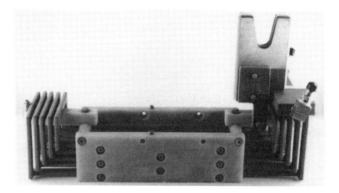

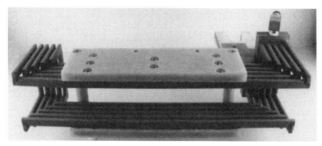

Fig. 4.28 Multi-turn channel coil. Source: Ref 12

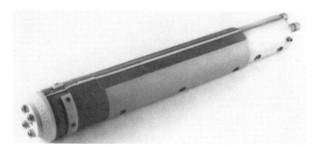

Fig. 4.29 Single-turn bore coil. Source: Ref 12

- *Fig. 4.30:* Multi-turn tube coil wound around Ferrotron flux intensifier designed for ID annealing a threaded hole after the workpiece was first carburized
- *Fig. 4.31:* Multi-turn rectangular tube coil wound on rigid support designed for ID hardening of an internal spline
- *Fig. 4.32:* Single-turn machined coil with integral quench, rigid mounting, and with a shorting slide for tuning (trombone) designed for internal hardening of a bearing. The trombone adjusts the coil impedance to help with tuning for output power.

Fig. 4.30 Multi-turn inside diameter (ID) coil with concentrator. Source: Ref 12

Fig. 4.31 Multi-turn coil rectangular tubing. Source: Ref 12

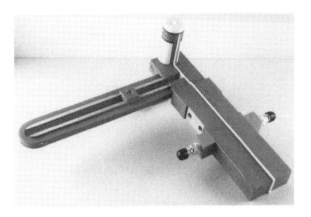

Fig. 4.32 Single-turn bore coil with trombone. Source: Ref 12

Pancake coils are wound in a circular manner that resembles the heating elements on a stove and are used to produce round or oblong patterns on flat surfaces. If a full circular pattern is needed, there will be a dead spot in the center of the coil unless a flux concentrator is used, or unless the workpiece is rotated off center or oscillated. Figure 4.33 shows a two-turn pancake coil designed for heating the edges of a thin plate. The pancake coils can be wound so that the concentric turns produce a complete spiral pattern, with the center lead extending upward out of the coil.

Butterfly Coil. Figure 4.34 shows an example of a butterfly coil designed for brazing a fitting to a tube. This is wound so that the outside vertical turns, which produce the heating, have their current flow in the

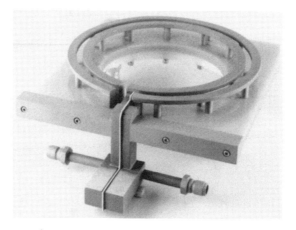

Fig. 4.33 Two-turn pancake coil. Source: Ref 12

Fig. 4.34 Butterfly coil. Source: Ref 12

same direction. This coil is an example of a transverse flux coil, and it is used to heat flats and thin cross sections. Because the induced current flows in the same direction, there is no dead spot as found with a pancake coil.

Split-Return Coil. Figure 4.35 shows two split-return coils. The coil is designed so that it has three leads. Two leads on the outside are parallel with each lead passing half the current, and the lead in the center is in-series carrying the full return current. Split-return coils are used for hardening parts such as the ends of shafts or bolts where a channel type coil would not heat the end face. Figure 4.35(a) a shows a split return with the return offset, providing a large window for the workpiece that was designed for hardening a groove. Figure 4.35(b) shows a split return with the return in the center and flux concentrator between the turns; quench

(a)

(b)

Fig. 4.35 Split return coil. Source: Ref 12

pads are on the outside. This coil was designed for hardening a lobe on a camshaft.

Reverse-Turn Coil. Figure 4.36(a) shows two coil designs that use the same coil to heat two different areas of the same workpiece. In Fig. 4.36(b), the design shows the coil wound so that the current travels in both sets of turns in the same direction. The design in Fig. 4.36(c) shows the upper turns with the current traveling in the opposite direction to the lower turns. The flux density between the two groups of turns increases as the spacing

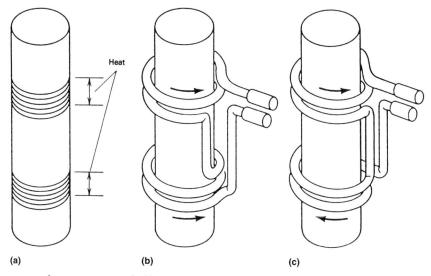

Fig. 4.36 Control of heating patterns in two different regions of a workpiece by winding the turns in opposite directions. Source: Ref 3

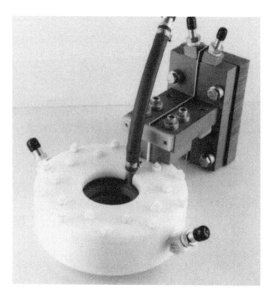

Fig. 4.37 Two-turn coil with opposing leads. Source: Ref 12

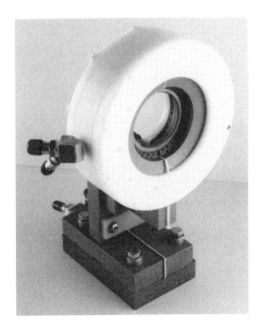

Fig. 4.37 (continued) Two-turn coil with opposing leads. Source: Ref 12

decreases, producing less heat between the two groups. When static heating cycles, this effect can be used to keep shoulders of parts from overheating where there are changes in diameter between the two areas. Figure 4.37 shows an example of a split return with opposing loops. A single-turn loop is at the top of the coil, with three turns of the opposing loop down inside the coil. The coil quenches through the flux concentrator.

Special Coils. Virtually any of the above coil configurations can be wound into special shapes to heat specific areas of parts. Figure 4.38

Fig. 4.38 Longitudinal heating coil, single shot with quench. Source: Ref 12

shows a single-shot coil oriented for longitudinal, single-shot hardening. Figure 4.39 shows an example of a coil built for single-tooth gear hardening by scanning. The coil is profiled to the pitch of the gear tooth, and it hardens one side of a tooth to the root, and back up the side of the adjacent tooth. Figure 4.40 shows a multiturn solenoid type coil used in a dual frequency heating application. Note the offset in the tubing so that the turns are parallel to the stud mounting for rigidity.

Bus Bars for Power Transmission. The bus bars used for power transmission from the terminal of the output transformer to the connection point of the coil need to be built rigidly and to have adequate cooling. Wide bus bars with close spacing between them will prevent power losses. Figure 4.41 provides examples of two bus bars. One is jogged so that the coil connection point is raised. This would be used in applications such as the bore hardening of cylinder lines in which the coil needs a high locating point so that it can extend down into the liner when scanning. The adjustable Y-axis bus bar permits adjustment of the coil height on the y-axis.

Fig. 4.39 Single-tooth gear coil. Source: Ref 12

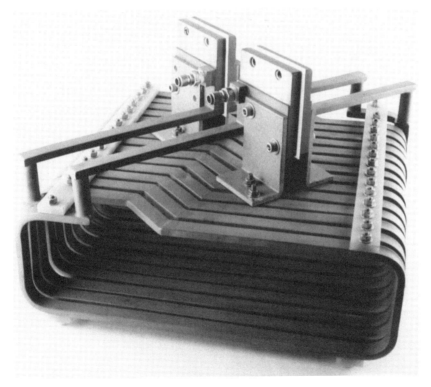

Fig. 4.40 Multi-turn solenoid coil with dual frequency. Source: Ref 12

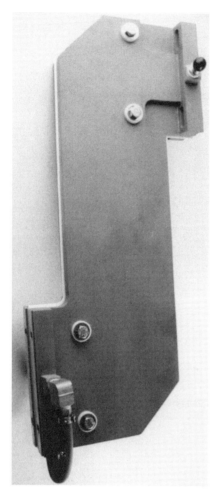
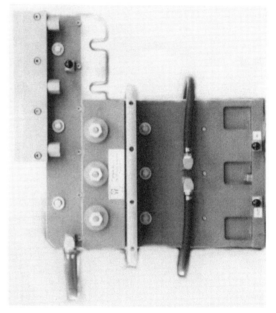

Fig. 4.41 Bus bar extensions. Source: Ref 12

Quick Connects. All coils need to be attached to the high-frequency output of the power supply, usually a transformer or copper bus bar connection. Historically this means of attachment has required the use of a mating connecting point so that each coil locates the same way (such as the bolt holes found in the Jackson type transformer) and the use of cooling water interconnection points. Quick connecting coil terminals have been developed to permit relatively simple and fast methods of connecting a coil. Figure 4.42 shows examples of several quick connects for RF while Figure 4.43 shows a quick connect for medium or low frequency.

Tempering Coils. Where the workpieces are magnetic, induction tempering is performed at temperatures below the Curie. The heating rate is slower, and possible heating should be done from the opposite side of the area to be tempered. Coils can be coupled more loosely, and in order to provide deeper heating, lower frequencies are used than would be used

Induction Coils / 71

Fig. 4.42 Adapters with coils. Source: Ref 12

Fig. 4.43 Dual bus bar with coils. Source: Ref 12

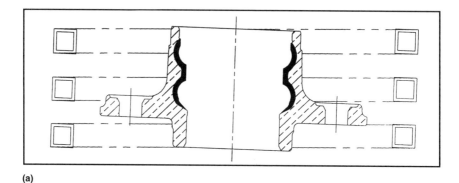

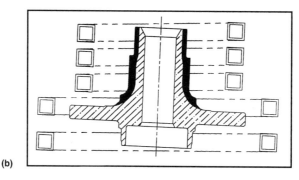

Fig. 4.44 Induction tempering of a part that has been (a) hardened on the inside surface and (b) hardened on the outside surface. Source: Ref 13

with austenitizing, Fig. 4.44 illustrates coil returns for the induction tempering of wheel spindles and wheel hubs. The wheel spindle is tempered from the same side as the case because internal heating is too difficult. If possible, however, the wheel hub can be heated from the opposite side of the of the area to be tempered. Induction tempering is discussed in the Chapter "Tempering" in this book.

REFERENCES

1. *Heat Treating*, Vol 4, *Metals Handbook*, 9th ed., American Society for Metals, 1981
2. Ajax Magnathermic, unpublished data
3. *Heat Treating*, Vol 4, *ASM Handbook*, ASM International, 1991
4. S.L. Semiatin and S. Zinn, *Induction Heat Treating*, ASM International, 1988
5. R.E. Haimbaugh, Induction Heat Treating Corp., personal research
6. V. Rudnev et al., Magnetic Flux Concentrators, *Met. Heat Treat.*, March/April 1995
7. R.S. Rufini et al., Advanced Design of Induction Heat Treating Coils, *Ind. Heat.*, Nov 1998

8. T. Learman, Formable and Thermal Hardenable Concentrator Enhances Induction Heating Process, *Ind. Heat.*, June 1995
9. G. Gariglio, "Monitoring and Controlling the Induction Heat Treat Process," paper presented at Furnace North America Show (Dearborn, MI), Sept 1996
10. V. Rudnev et al., *Steel Heat Treatment Handbook*, Marcel Dekker, Inc., 1997
11. S.L. Semiatin and D.E. Stutz, *Induction Heat Treating of Steel*, American Society for Metals, 1986
12. Induction Tooling, unpublished data
13. K. Weiss, Induction Tempering of Steel, *Adv. Mater. Process.*, Aug 1999

CHAPTER 5

Heat Treating Basics

THE INTENT OF this Chapter is to provide the basic metallurgical theory necessary to understand the induction heat treating practice. Heat treating is the treatment of metals by heating and cooling in a prescribed manner so as to obtain specific conditions and/or properties. Almost all metals and metal alloys respond to some form of heat treatment. While almost any pure metal or alloy can be softened by means of a suitable heating and cooling cycle, only certain alloys can be strengthened or hardened by heat treatment. The reason that steels account for more than 80% of total metal production is that practically all steels respond to one or more types of heat treatment. The heat treatment of a specific steel part may involve several different types of heating and cooling processes in sequence. The heating can be accomplished in a wide variety of ways, ranging from the use of heat treating furnaces to lasers, while cooling methods can range from cooling in air to quenching with water. Regardless of the method of heating and cooling used, metallurgical processes can be described scientifically.

Iron and Steel

Fundamentally, all steels are alloys of iron and carbon with carbon as the principal ingredient. The amount of carbon present in steel has a pronounced effect on the properties of a steel and on the selection of suitable heat treatments to produce desired properties. So-called plain carbon steels have low amounts of carbon while also having small but specified amounts or percentages of manganese and silicon, as well as small unavoidable but measurable and limited amounts of phosphorus and sulfur. In all carbon steels, small quantities of alloying elements or residual elements such as nickel, chromium, and molybdenum are present. These residual elements are unavoidable because they are retained from the raw materials used in melting. However, as a rule, small amounts of these elements have little or no disadvantage to the intended use. To impart special characteristics or properties for engineering applications, one or more of several other elements may be added to alloy steel.

Standard carbon and alloy grades established by the American Iron and Steel Institute (AISI) or the Society of Automotive Engineers (SAE) have now been assigned designations in the Unified Numbering System (UNS) by ASTM (ASTM E 527) and SAE (SAE J 1086). There are a number of different systems, although the AISI numbering system is the most commonly accepted. Table 5.1 shows the types and approximate percentages of identifying elements in standard carbon and alloy steels that are induction hardened. Most numbers are four digits long, with a few, such as AISI 52100, being five digits long. The last two digits of the carbon steel indicate the carbon content. Most induction-hardened parts in the four digit series will have a carbon content under 0.60%. Other U.S. numbers such as ASTM and SAE designations may be used, and there are various international designations with the AISI correlation. A general cross-reference to the other types of designations is listed in of Ref 2.

Carbon Steel. Carbon steels, referred to as plain carbon steels, are classified in four distinct series in accordance with the AISI system. All four series can be discussed in terms of their carbon content in general response to heat treating:

- The 1000 series are plain carbon steels containing not more than 1.00% manganese maximum.
- The 1100 series are resulfurized carbon steels for machinability (sulphur is added).
- The 1200 series are resulfurized and rephosphorized carbon steels.
- The 1500 series have up to 1.65% manganese.

Plain carbon steels are divided into three groups depending upon carbon content.

Group I—Low Carbon (0.08 to 0.25% Carbon). The three principal types of heat treatment used on these low-carbon steels are (a) process treating of material to prepare it for subsequent operations; (b) treating of finished parts to improve mechanical properties; and (c) case hardening by carburizing or carbonitriding to develop a hard, wear-resistant surface. Some parts containing more than 0.18% carbon are induction hardened to improve wear and strength.

Group II—Medium Carbon (0.30 to 0.55% Carbon). Because of the higher carbon content, quenching and tempering become increasingly important when steels of this group are considered. They are the most versatile of the carbon steels because their hardenability (response to quenching) can be varied over a wide range by suitable controls. In this group of steel there is a continuous change from water-hardening to oil-hardening steels. The hardenability is very sensitive to changes in chemical composi-

Table 5.1 Classification of carbon and steel alloys

AISI Designation	Description C	Composition, %				
		Mn	Cr	Mo	Ni	Va
Carbon Steels						
10xx	xx	0.30–1.50	...	...	...	...
11xx	xx	1.30–1.65	...	...	...	...
15xx	xx	1.25–1.75	...	...	...	...
Alloy Steels						
41xx	xx	0.40–1.00	0.80–1.10	0.15–0.25	...	...
43xx	xx	0.45–0.80	0.40–0.90	0.20–0.30	1.65–2.00	...
52100	0.98–1.10	0.25–0.45	1.30–1.60	...	...	...
6150	0.48–0.53	0.70–0.90	0.80–1.10	...	...	0.015 min.
H-Steels						
4150H	0.65–1.10	0.65–1.10	0.75–1.20	0.15–0.25	...	...
Boron Steels						
10Bxx	xx					
15B35	0.31–0.39	0.70–1.20	...	...	...	...
Stainless Steels						
416	0.15 max	1.25 max	12.0–14.0	...	...	...
420	Over 0.15	1.00 max	12.0–14.0	...	...	...
440C	0.96–1.20	1.00 max	16.0–18.0	0.75 max	...	...

The xx in the last two digits stands for the carbon content by weight. Steel 1045 has 0.45% carbon. Source: Ref 24

tion, particularly to the content of manganese, silicon, and residual elements, as well as grain size. These steels are also very sensitive to changes in cross section.

Carbon steels may be purchased as modified steels in which small amounts of specific elements, such as manganese, are added to increase the hardenability. There is also widespread use of the resulfurized grades such as AISI 1141 and 1144, which are readily induction hardened. Some of these grades are cold drawn at room temperature or at elevated temperatures using heavier-than-normal drafts with a following stress relief and are used in place of quenched and tempered bars. The heavier drafts produce higher tensile and yield strengths. While parts made from bar stock are generally given no prior heat treatment before induction hardening, the medium-carbon steels need homogeneous microstructures for the best response to the fast austenitizing cycles. Heat treatment for the refinement of the microstructure is often desirable.

Group III—High Carbon (0.60 to 0.95% Carbon). Oil or polymer quenches are used to minimize distortion and the potential for quench cracking. As with the medium carbon steels, prior heat treatment may be necessary to provide a microstructure that will give the best response with induction.

Alloy steels are those steels containing specified percentages of other elements in their chemical compositions. The elements most commonly alloyed with steel are nickel, chromium, molybdenum, vanadium, and tungsten. Manganese also falls into this category when it is specified in percentages generally in excess of 1%. As mentioned in the Chapter "Heat Treating of Metal" in this book, induction heating came into prominence because of its ability to use carbon steels for the replacement of alloy steels. Alloy steels are not normally used unless they are first quenched and tempered to provide core hardness, or the hardenability is needed for deep case depths. Hardenability will be discussed later in this Chapter.

H-Steels. H-steels are guaranteed by the supplier to meet established hardenability limits for specific ranges of chemical composition. An example of designation of an H-steel would be 1045 H. This is 1045 steel warranted for hardenability. When an H-steel is specified, the steel producer shows on the shipping papers or by some other acceptable means the hardenability characteristics of the heat of steel involved.

Boron Steel. Boron is a potent and economical alloying element that markedly increases hardenability when added to fully deoxidized steel. The effects of boron on hardenability are unique in several respects:

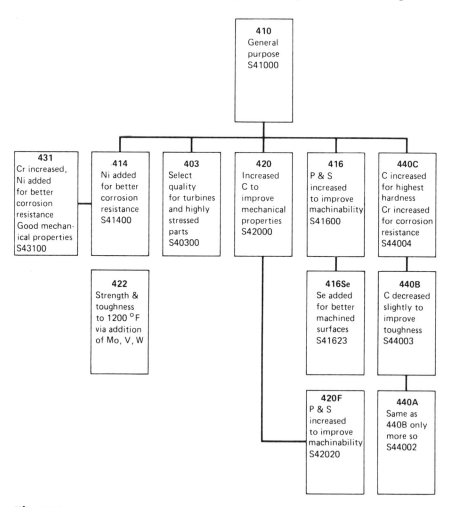

Fig. 5.1 Family relationships for standard martensitic steels. Source: Ref 3

(a) a very small amount of boron (about 0.001%) has a powerful effect on hardenability; (b) the effect of boron on hardenability is much less in high-carbon than low-carbon steels. Very small amounts of boron are sometimes added to the medium-carbon steels to increase their hardenability.

Stainless Steel. Stainless steels are best defined as alloy steels containing at least 10% chromium, with or without other elements. Chromium is added to increase corrosion resistance. Stainless steels are commonly divided into four groups, of which only the martensitic family is induction hardened. The martensitic grades have AISI classifications composed of three numbers or three numbers and a letter. The numbers are all in the 400 series, but not all 400 series grade designations are martensitic. (Some are ferritic.) The grades most commonly induction hardened are 416, 420, 440C, and 440F. Figure 5.1 shows the family relationships for standard martensitic stainless steels.

Cast Iron. The term *cast iron*, like the term *steel*, identifies a large family of ferrous alloys. Cast irons primarily are alloys of iron that contain more than 2% carbon and from 1 to 3% silicon. Cast irons have more carbon present than can be retained in solution in solid metal. The carbon is found both combined with iron as iron carbide (cementite) and as graphite (one of the crystalline forms of carbon). Wide variations in properties can be achieved by varying the balance between carbon and silicon, by alloying with various metallic or nonmetallic elements, and by varying melting, casting, and heat treating practices. Table 5.2 shows the general classification of cast irons by tensile strength according to ASTM class numbers. There are six basic types of cast irons and several varieties of each type. The types of iron are classified by how excess carbon occurs in the microstructures. Figure 5.2 shows the types of cast iron in relationship to the carbon and silicon content. The three types of cast iron most commonly induction hardened are malleable iron, gray iron, and ductile iron. Gray and ductile irons are sometimes confused.

Table 5.2 Classification of cast irons by tensile strength

ASTM class	Approximate tensile strength	
	MPa	ksi
20	138	20
25	172	25
30	207	30
35	241	35
40	276	40
45	310	45
50	345	50
60	414	60

Source: Ref 1

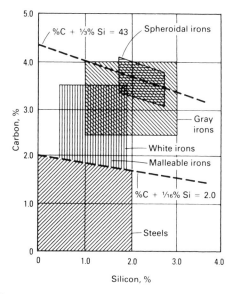

Fig. 5.2 Cast iron in relationship to carbon and silicon. Source: Ref 4

Identification is important in heat treating because of the typically lower silicon content and wider alloy content in the higher strength grades of gray iron.

Malleable iron contains compact nodules of graphite flakes in the form known as temper carbon. As opposed to the spheroidal nodules found in ductile cast iron, these nodules are somewhat irregular aggregates. Numbers such as 40010, as shown in Table 5.3, classify the pearlitic and martensitic ASTM grades that can be induction hardened.

Gray iron contains carbon in the form of graphite flakes. Gray cast irons are classified by ASTM specification A48, which classifies the various types in terms of tensile strength. The ASTM class numbers range from Class 20 to Class 60

Ductile iron, also known as spheroidal graphite iron, contains spherulitic graphite in which the graphite flakes form into balls. Ductile iron is so named because in the as-cast form it exhibits measurable ductility.

Table 5.3 Classification of malleable iron by mechanical properties

ASTM A220 class or grade	Hardness, HBN	Tensile strength MPa	ksi
40010	149–197	414	60
45008	156–197	448	65
45006	156–207	448	65
50005	179–229	483	70
60004	197–241	552	80
70003	217–269	586	85
80002	241–285	655	95
90001	269–321	724	105

Source: Ref 1

Table 5.4 Average mechanical properties of ductile cast iron

Nearest standard grade	Hardness HB	Ultimate strength	
		MPa	ksi
60-40-18	167	461	66.9
65-45-12	167	464	67.3
80-55-06	192	559	81.1
120-90-02	331	974	141.3

Source: Ref 1

Most of the specifications for standard grades of ductile iron are based on properties—that is, strength and/or hardness is specified for each grade. Composition is either loosely specified or made subordinate to the mechanical properties. Table 5.4 shows the classification and numbering system according to average mechanical properties. Note that the classifications are in three segmented numbers such as 60-40-18.

Tool steels are any steels used to make tools for cutting, forming, or otherwise shaping a material into a part or component adapted to a definite use. The earliest tool steels were simple, plain carbon steels. Beginning in 1868 and, to a greater extent, early in the 20th century, many complex, highly alloyed tool steels were developed. These complex alloy tool steels contain relatively large amounts of other elements such as tungsten, molybdenum, manganese, and chromium. Very precise production practices and stringent quality controls are used to fabricate tool steels. With few exceptions, all tool steels must be heat treated to develop specific combinations of physical and mechanical properties. The induction hardening of tool steels is limited to a few types of tool steels and specific, uniform geometry. While induction is used to heat some tool steels, the complex shapes and heating requirements for most tools and dies do not represent a good application for induction. Table 5.5 shows the classifications of tool steels that are most commonly induction hardened. Induction tempering can be done on some tool steels for slight reduction of the as-quenched hardness to help increase ductility.

Powdered metal parts are formed by compressing precisely mixed metal powders of iron and desired alloying elements in a shaped die to produce what is termed a *green* compact. The compacts are then sintered (diffusion bonded) at elevated temperature in a furnace with a protective atmosphere. During sintering, the powdered metal is effectively fused together. By pressing and sintering only, parts of more than 80% theoretical density of iron can be produced. By repressing, the density can be increased to 90%.

Iron, iron-copper, iron-copper-carbon and prealloyed steels are among the materials used for powder metallurgy parts. Parts made from many of these materials respond to heat treatment. Powder metallurgy materials customarily are designated by the specifications or standards to which

Table 5.5 Tool steel classification by major elements

AISI Class	Composition, %								
	C	Mn	Si	Cr	Ni	Mo	W	V	Co
Air-hardening, medium alloy cold work steels									
A2	0.95–1.05	1.00 max	...	4.75–5.50	...	0.90–1.40	...	0.15–0.50	...
A3	1.20–1.30	0.40–0.60	...	4.75–5.50	...	0.90–1.40	...	0.80–1.40	...
A4	0.95–1.05	1.80–2.20	...	0.90–1.20	...	0.90–1.40	...	...	...
A6	0.65–0.75	1.80–2.50	...	0.90–1.20	...	0.90–1.40	...	...	...
A7	2.00–2.85	0.80 max	...	5.00–5.75	...	0.90–1.40	0.50–1.50	3.90–5.15	...
A8	0.50–0.50	0.050 max	0.65–1.10	4.75–5.50	...	1.15–1.65	1.00–1.50	...	...
A9	0.45–0.55	0.050 max	0.95–1.15	4.75–5.50	1.25–1.75	1.30–1.80	...	0.80–1.40	...
A10	1.25–1.50	1.60–2.10	1.00–1.50	...	1.55–2.05	1.25–1.75	...	...	...
High-carbon, high-chromium cold work steels									
D2	1.40–1.60	0.60 max	...	11.00–13.00	...	0.70–1.20	...	1.00 max	1.00 max
D3	2.00–2.35	0.60 max	...	11.00–13.00	1.00 max	...	1.00 max	...	...
D4	2.05–2.40	0.60 max	...	11.00–13.00	...	0.70–1.20	...	...	...
D5	1.40–1.60	0.60 max	...	11.00–13.00	...	0.70–1.20	...	2.50–3.50	2.50–3.50
D7	2.15–2.50	0.60 max	...	11.50–13.50	...	0.70–1.20	...	...	...
Oil-hardening cold work steels									
O1	0.85–1.00	1.00–1.40		...	0.40–0.60	...	0.40–0.60	...	...
O2	0.85–0.95	1.40–1.80		...	...	...	...	...	...
O6	1.25–1.55	0.30–1.10		...	...	...	...	...	...
O7	1.10–1.30	1.00 max		...	1.00–2.00	...	1.00–2.00	...	...
Shock-resisting steels									
S1	0.40–0.55	0.10–0.40	0.15–1.20	1.00–1.80	1.50–3.00	0.050 max	1.50–3.00	...	...
S2	0.40–0.55	0.30–0.50	0.30–0.50	...	...	0.30–0.60	...	...	...
S5	0.50–0.65	0.60–1.00	0.60–1.00	...	...	0.20–1.35	...	...	...
S6	0.40–0.50	1.20–1.50	1.20–1.50	1.20–1.50	...	0.30–0.50	...	...	...
S7	0.40–0.55	0.20–0.80	0.20–0.80	3.00–3.50	...	1.30–1.80	...	...	...

"..." indicates only present as a very small amount or residual amount. Source: Ref 1

they are made, such as those listed in Table 5.6. The ASTM and SAE designations use the term *class* to indicate the carbon content of the finished part. The term *grade* is used to differentiate between related alloys bearing the same designation. Metal Powder Industries Federation (MPIF) designations for ferrous powder metallurgy materials include a prefix of one or more letters (the first of which is "F" to indicate an iron-based material), four numerals, and a single-letter suffix.

Iron/Iron-Carbide

Carbon in Iron. Pure iron by itself is very limited in engineering usefulness. Carbon is the main alloying element in the different crystalline forms of iron that are called steel, and it produces the superior physical properties that make steel so useful. Even in the highly alloyed stainless steel, it is the quite minor constituent carbon that virtually controls the engineering properties. Furthermore, due to the manufacturing processes, carbon in effective quantities persists in all irons and steels unless special methods are used to minimize it. Iron alloyed with carbon forms the phase called *iron carbide,* commonly called *cementite* in metallurgical terms. In

Table 5.6 Ferrous powdered metal compositions

Description	MPIF Class	MPIF compostion ranges, %			
		C	Ni	Cu	Fe
Iron	F-0000	0.30 max	...	...	97.70–100.00
Steel	F-0005	0.30–0.60	...	...	97.40–99.70
Steel	F-0008	0.60–1.00	...	...	97.00–99.10
Copper iron	FC-0200	0.30 max	...	1.50–3.90	93.80–98.50
Copper steel	FC-0205	0.30–0.60	...	1.50–3.90	93.50–98.20
Copper steel	FC-0208	0.60–1.00	...	1.50–3.90	93.10–97.90
Copper steel	FC-0505	0.30–0.60	...	4.00–6.00	91.40–95.70
Copper steel	FC-0508	0.60–1.00	...	4.00–6.00	91.00–95.40
Copper steel	FC-0808	0.60–0.90	...	6.00–11.00	86.00–93.40
Nickel steel	FN-0205	0.30–0.60	1.00–3.00	2.50 max	91.90–98.70
Nickel steel	FN-0208	0.60–0.90	1.00–3.00	2.50 max	91.60–98.40
Nickel steel	FN-0405	0.30–0.60	3.00–3.50	2.00 max	89.90–96.70
Nickel steel	FN-0408	0.60–0.90	3.00–3.50	2.00 max	89.60–96.40
Nickel steel	FN-0705	0.30–0.60	6.00–8.00	2.00 max	87.40–93.70
Nickel steel	FN-0708	0.60–0.90	6.00–8.00	2.00 max	87.10–93.40
Infiltrated steel	FX-1005	0.30–0.60	...	8.00–14.00	80.50–91.70
Infiltrated steel	FX-1008	0.60–1.00	...	8.00–14.00	80.10–91.40
Infiltrated steel	FX-2005	0.30–0.60	...	15.00–25.00	70.40–84.70
Infiltrated steel	FX-2008	0.60–1.00	...	15.00–25.00	70.00–84.40

MPIF, the Metal Powder Industries Federation. Source: Ref 5

induction heat treating practice, it is common to use iron carbide or carbide instead of cementite.

Carbon in iron has the capability of forming a solid state solution that makes possible the successful heat treating of steel. Carbon atoms are one-thirtieth the size of iron atoms. Carbon is almost insoluble in the ferritic stage, but very soluble in the austenitic phase that exists above 725 °C (1340 °F). Carbon actually dissolves; that is, the individual atoms of carbon lose themselves in the interstices among the iron atoms. The solid solution of carbon in iron can be visualized as a pyramidal stack of basketballs with golf balls between the spaces in the pile. In this analogy the basketballs would be the iron atoms, while the golf balls interspersed between them would be the smaller carbon atoms. The austenitic phase is much more accommodating to carbon than the ferritic phase, with the explanation of this based on fundamental theory involving electron bonding and the balance of the attractive and repulsive forces.

Phases of Iron-Carbon. A phase is a portion of an alloy—physically, chemically, or crystallographically homogeneous throughout—that is separated from the rest of the alloy by distinct boundary surfaces. The following phases occur in iron-carbon alloys: molten alloy, austenite, ferrite, cementite, and graphite. Please note that any of these phases may likewise be called constituents (ingredients in a system). Therefore, not all constituents such as pearlite and bainite are phases because they are mixtures of phases and not homogeneous throughout. A phase diagram is a graphical representation of the equilibrium temperature and composition limits of phase fields and phase reactions in an alloy system. Figure 5.3 shows the iron-cementite phase diagram. Phase diagrams have the temperature plotted vertically and composition (percent carbon in iron), horizontally.

The left side of the diagram is 100% iron. Proceeding from left to right, carbon increase from 0 to 6.67%. Phase diagrams should be viewed mainly from an equilibrium viewpoint. The phases, as shown, consider that time is not a factor and that the phases exist at the temperature shown. This diagram is very important in understanding the basic nature of the phases and is important in understanding the beginning basic theory of heat-treating.

The iron-cementite diagram is usually called the iron-carbon diagram. Technically this is incorrect because iron-carbide is the phase called cementite. Since cementite under equilibrium conditions decomposes into iron and graphite, the technically correct diagram is that shown in Fig. 5.4: the iron-graphite diagram. This diagram has significance only in the heat treating of some cast irons and powdered metal parts where graphite is present.

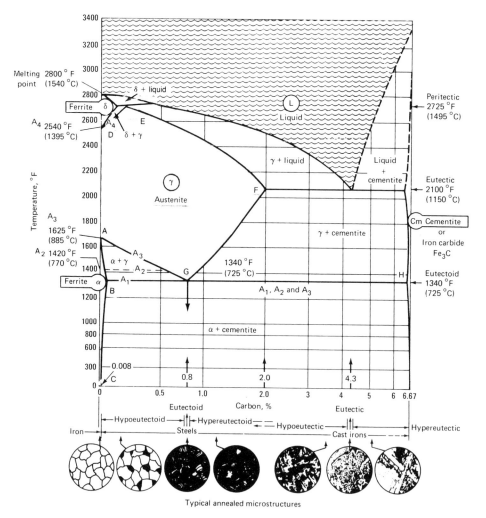

Fig. 5.3 Iron-cementite diagram. Source: Ref 2

The process by which iron changes from one atomic arrangement to another when heated or cooled is called a *transformation*. Transformations that occur at very slow rates of heating and cooling are known as *equilibrium transformations*. Transformations of this type occur not only in pure iron, but also in many of its alloys. Ferritic iron alloyed with carbon is transformed to austenitic iron above the upper transformation temperature. It is this transformation that makes possible the variety of properties that can be achieved to a high degree of reproducibility through the use of carefully selected heat treatments.

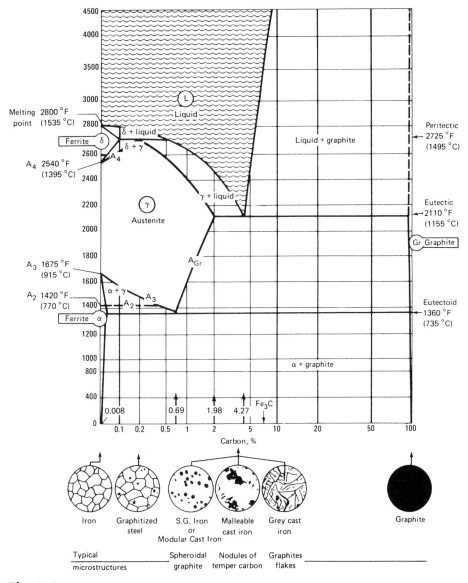

Fig. 5.4 Iron-graphite diagram. Source: Ref 2

When carbon atoms are present in iron, two changes occur. First, depending on the carbon content, transformation temperatures are changed; and second, transformation takes place over a range of temperatures rather than at a single temperature. Although the iron-cementite phase diagram as shown in Figure 5.3 extends from a temperature of 1870 °C (3400 °F) down to room temperature, note the part of the diagram that lies below 1121 °C (2050 °F). Steel induction heat treating practice rarely involves the use of temperatures above 1121 °C (2050 °F).

Austenite is the term applied to the phase that is a solid-state solution of carbon in iron. The solubility of carbon in austenite depends on temperature and can range from 0 to 2%. Under normal equilibrium conditions austenite cannot exist in carbon steel at room temperature. (It may be present in alloy and stainless steels.)

Classification of Heat Treating Processes

The intent of this book is to cover induction heat treating practices. In some instances, heat treatment procedures are clear-cut in terms of technique and application. In other instances, descriptions or simple explanations are insufficient because the same technique frequently may be used to obtain different objectives. For example, the processes of stress relieving and tempering are often accomplished with the same equipment and by using the identical time and temperature cycles. The objectives, however, are different for the two processes.

It is important that the heat treater understand the general classifications of heat treating processes and their importance before induction hardening, and where these processes fit into the practice of induction heat treating. The following descriptions of the principal heat treating processes are arranged according to their interrelationships. The processes are discussed as to their relationship with induction heat treating, along with the metallurgical aspects.

Annealing is a generic term denoting a heat treatment that consists of heating to and holding at a suitable temperature, followed by cooling at a suitable rate. The cooling rates at high temperature are slow, and the cycle during cooling may involve holding times at elevated temperatures. Cycles are used to produce a microstructure that provides high softness and high machinability. The microstructures in steel obtained by annealing are not generally desirable for induction heating.

When the term *induction anneal* is used, it generally means to soften the metal by heating with induction. Because there is no holding at temperature time for induction, steel parts that are induction annealed are in reality induction tempered (see "Tempering" in this Chapter). Induction is used to anneal nonferrous metals or alloys such as copper or brass.

However, with nonferrous alloys, the annealing cycles involve heating the workpiece to full solution and quenching fast enough to anneal. Electrical contacts made from beryllium copper are solution hardened in the furnace and induction annealed on the barrel ends so that the ends can be crimped.

Normalizing consists of heating a ferrous alloy to a suitable temperature above its specific upper transformation temperature and holding at temperature for time sufficient to complete transformation. Cooling at a predetermined cooling rate and cycle follows this. The cooling rate is usually faster than that used for annealing, and there is not generally a holding time at elevated temperature during the cooling cycle. Normalizing is usually used as a conditioning treatment for strength, machinability, or to produce a microstructure that is better for induction hardening.

Austenitizing is defined as the process of forming austenite by heating steel above the transformation range. This term will be widely used in the remaining chapters of this book. When used without qualification, the term implies complete austenitizing. Workpieces that are to be induction hardened are first austenitized. Austenitizing is also the initial step prior to normalizing, full annealing, and quench hardening of ferrous alloys.

Quenching is the rapid cooling of a steel or alloy from the austenitizing temperature through contact with a quenching medium. Induction heat treating uses dunk immersion into a quench tank, spray quenches, and mass quenching. Most common quenching media include water, water-polymer solutions, and oil. When selecting a quenchant, in order to minimize the possibility of quench cracks and distortion, it is best to avoid a solution that has more cooling power than is needed to achieve the results. Selection of the quenchant may require a complete analysis of all the factors involved in the induction heat treating practice. The Chapter "Quenching" in this book discusses quenching practices, and the Chapter "Induction Heat Treating Process Analysis" presents a method for process analysis.

Tempering consists of the reheating of ferrous alloys to some preselected temperature that is below the lower transformation temperature. Tempering is used to increase the ductility of heat treated parts. Most parts are tempered from 107 to 538 °C (225 to 1000 °F). Parts may be furnace tempered or induction tempered. The Chapter "Tempering" in this book discusses tempering.

Stress relieving for steels and irons, like tempering, is always done by heating to some temperature below the transformation temperature. The primary purpose of stress relieving is to relieve (make uniform throughout the cross section) mechanical stresses that have been imparted to parts from such processes as forming, rolling, machining, welding, or heat treating. Stress relieving is normally a batch, furnace-type of heat treatment. The parts are heated and are held at the established temperature only long enough to reduce the residual stresses to an acceptable level. Then the parts are cooled at a relatively slow rate to avoid creation of new stresses.

Stress relieving is normally performed at higher temperature ranges, such as from 538 to 650 °C (1000 to 1200 °F). However, the term is also used where hardened parts are heated at the 120 to 150 °C (250 to 300 °F) range for purposes of making the parts more ductile without causing a decrease in hardness.

Carburizing consists of absorption and diffusion of carbon into low carbon steels or alloys by heating to some temperature above the upper transformation temperature of the specific alloy. Temperatures used for carburizing are generally in the range of 900 to 1040 °C (1650 to 1900 °F). Heating is done in a carbonaceous environment (liquid, solid, or gas). This raises the carbon content at the surface, producing a carbon gradient extending inward from the surface. The surface layers can then be hardened to a high hardness either by quenching from the carburizing temperature or by cooling to room temperature, followed by a re-austenitizing and quenching. Parts that have been carburized can be induction tempered in localized areas, and parts that have been carburized and slow cooled or tempered can be induction hardened in localized areas.

Carbon restoration is a process in which gas carburizing is used to restore the decarburized surface of a steel or alloy part. The carbon potential is controlled so that the carbon content of the decarburized surface is increased to the same content as or slightly higher content than the carbon content of the part. For instance, an AISI 1050 steel with surface decarburization may have the carbon restored 0.50%. (See discussion of decarburization in the Chapter "Decarburization and Defects" in this book.)

Carbonitriding is a case-hardening process in which a ferrous alloy, generally low carbon steel, is heated above the transformation temperature in a gaseous atmosphere of such composition as to cause simultaneous absorption of carbon and nitrogen on the surface. This process is used to produce cases that are shallower than those produced by carburizing. Thin, hard, wear-resistant cases are produced because of the formation of complex carbides and nitrides in the case. The nitrides can be resistant to induction tempering.

Nitriding is a shallow, case hardening process in which a ferrous alloy is heated at a temperature below the transformation temperature in a nitrogenous atmosphere, usually ammonia, generally in the 565 °C (1000 °F) region. Quenching is not required to create a hard, wear-resistant and heat-resistant case. Alloy steels, such as the AISI 4000 series steels, may be used. To produce the benefits of the induction-hardened case combined with the high hardness benefits of the nitrided case, parts may be induction hardened under nitrided cases, with no effect on the nitrided case.

Nitrocarburizing is a very shallow, case-hardening process done through contact with both nitrogen and carbon at temperatures below the transformation temperature. The purpose is to provide a low distortion method for producing shallow cases that improves the antiscuffing characteristics of ferrous parts.

Hardness and Hardenability

There are several methods for measuring or evaluating the results of the physical properties produced as a result of heat treating, such as tension testing, impact testing, bend testing, and various types of shear testing. However, most of these tests require special equipment and destruction of the part being tested, and the tests cannot be performed quickly in line. By far, hardness testing is the most universally used method for measuring the conformance to specifications for heat treating for several reasons:

- The hardness test is simple to perform.
- The hardness test does not usually impair the usefulness of the workpiece being tested.
- Accurately tested and properly interpreted, hardness tests correlate to other physical properties. Hardness varies directly with strength and inversely with ductility.

Hardness. Hardness is a matter of interpretation. To some people, a hard substance does not wear, while to others hardness is something that will not bend, that resists penetration, or that has the ability of an object to scratch (such as diamond on glass). The definition provided in most dictionaries is "the relative capacity of a substance for scratching another."

Many different systems have been devised for testing the hardness of metals, but only a few have achieved commercial importance. These are the Brinell, Rockwell, Vickers, rebound type, and various microhardness testers. Except for the rebound type, all of these testers depend on the principle of indentation for measurement. In practice, induction heat treating hardness testing uses Rockwell, Vickers, rebound types, and microhardness testers, with the Rockwell "C" test scale being the most versatile and widely used scale for specification of hardness. The Chapter "Standards and Inspection" in this book discusses in more detail the principles behind the different testers. Figure 5.5 shows a bench-type Rockwell hardness tester.

Hardenability is a term used widely throughout this book, and it is important that the reader understand its precise meaning. One might think that hardenability means the ability to be hardened. Although true to a degree, the precise meaning of hardenability is much broader. In classes on heat treating, two questions are asked: (a) a specific steel, after being heated above the upper critical temperature and then quenched in water, shows a hardness of 65 HRC (near the limits for most steels as measured by Rockwell testing); does this establish it as a high-hardenability steel? (b) after heating and quenching, a specific steel registers only 35 HRC; is it a low hardenability steel? Invariably a student will answer yes to the first question and no to the second. The correct answer for both is "not necessarily." In each instance, information on section thickness is required before either question can be correctly answered. Therefore, hardenability does not necessarily mean the ability to be hardened to a certain Rockwell or Brinell value.

For example, just because a given steel is capable of being hardened to 65 HRC does not necessarily mean that it has high hardenability. Also, a steel that can be hardened to only 35 HRC may have very high hardenability. Hardenability refers to the capacity of a steel to harden to a speci-

Fig. 5.5 Rockwell tester. Source: Ref 6

fied harden or hardness range at given depths. Deep hardening steels, such as alloy steels, are said to have high hardenability.

The Jominy Hardenability (end-quench) Test is one of the most commonly used hardenability tests for predicting the ability of induction heat treating to produce minimum specified case depths. In this test, test bars of normally 25.4 mm (1 in.) in diameter by 102 mm (4 in.) long are fully austenitized. The bars are then removed from the furnace and placed into the test fixture as shown in Fig. 5.6. Water flow from underneath is controlled so that the amount striking the end of the specimen is constant in volume and velocity. The water impinges on the end of the specimen only, cooling the part. Thus the end with quench impingement quenches the fastest. When the bar has completely cooled, the bar is removed from the fixture and parallel, opposite flats are ground so that the hardness can be checked down the longitudinal direction of the bar. Hardness tests are made every 1.5 mm (1/16 in.). Figure 5.7 shows an example of test results.

Workpieces that are induction hardened will generally have a minimum case depth specified. The hardenability curves are significant in that they present data that can be used to predict whether a given steel can meet specifications, providing the prior microstructure is present. Figure 5.8 shows the hardenability curve for 1045 steel. The hardness shown at the first 1/16 in. (J1), is representative of surface hardness, and the deeper hardnesses are representative of hardnesses that can be produced deeper with good quenching. If a workpiece specification called for the hardness to be HRC 64 at the surface or the hardness to be above 50 HRC at 6 mm (0.250 in.), the specification is not likely to be met. The hardenability

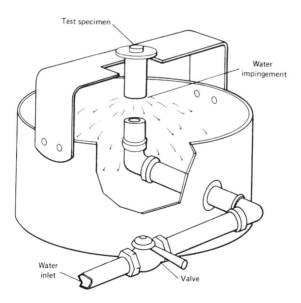

Fig. 5.6 Standard end-quench (Jominy) test specimen and method of quenching in quenching jig. Source: Ref 2

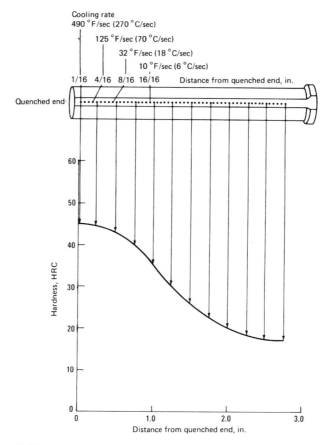

Fig. 5.7 Plotting of hardness vs distance for end quench. Source: Ref 2

charts of steel alloys commonly induction hardened appear in the Appendix "Hardenability Curves" in this book.

Role of Carbon in Hardened Steels. Carbon is the key to the hardening of steels by the heating and quick cooling (quenching) mechanism. The carbon content of a steel determines the maximum hardness attainable. The effect of carbon on attainable hardness is demonstrated in Fig. 5.9. The maximum attainable hardness requires only about 0.60% carbon, which probably seems odd to the reader. From the iron-cementite diagram (Fig. 5.3), it would seem logical that hardness would not become a straight line until about 0.77% C is reached. However, there is essentially no change in attainable hardness above about 0.60% C according to the data shown in Fig. 5.9. This can be accounted for through some unavoidable deficiencies concerned with indentation hardness testing. Despite this apparent oddity, these data are accurate and absolutely reproducible for extremely thin sections of carbon steel.

In fact, the data shown in Fig. 5.9 are precise to the extent that they sometimes are used in reverse. The hardness to which a specimen

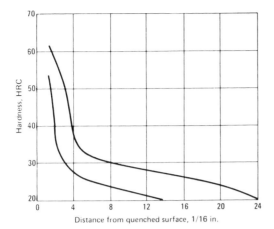

Distance from quenched surface		Hardness, HRC		Distance from quenched surface		Hardness, HRC	
1/16 in.	mm	max	min	1/16 in.	mm	max	min
1	1.58	62	55	7	11.06	31	25
1.5	2.37	61	52	7.5	11.85	30	24
2	3.16	59	42	8	12.64	30	24
2.5	3.95	56	34	9	14.22	29	23
3	4.74	52	31	10	15.80	29	22
3.5	5.53	46	29	12	18.96	28	21
4	6.32	38	28	14	22.12	27	20
4.5	7.11	34	27	16	25.28	26	...
5	7.90	33	26	18	28.44	25	...
5.5	8.69	32	26	20	31.60	23	...
6	9.48	32	25	22	34.76	22	...
6.5	10.27	31	25	24	37.92	21	...

Source: *Metals Handbook*, 9th ed., Vol 1, American Society for Metals, 1978

Fig. 5.8 End quench hardenability for 1045 steel. Source: Ref 2

quenches can be used as a quick means of determining carbon content of an unknown steel. For example, a properly austenitized and quenched workpiece that tests to low hardness during set up can be the first indication of low carbon content in the steel. In conventional heat treating practice, it should be remembered that the data shown in Fig. 5.9 are

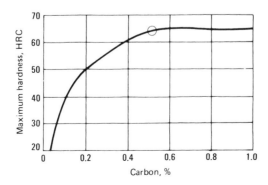

Fig. 5.9 Relationship between carbon content and hardness. Source: Ref 13

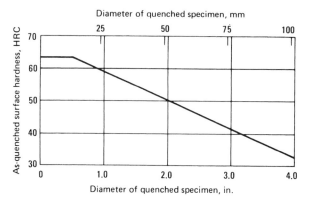

Fig. 5.10 Effect of section size on surface hardness of 0.54% carbon steel. Source: Ref 2

based on heat treating wafer-thin sections that are cooled from their austenitizing temperature to room temperature within a matter of seconds, thus developing 100% martensite. In practice the cross sections are usually larger, and less than optimum quenching speed may be used to avoid quench cracking. Also, the workpiece size may be so large that it cannot be quickly quenched.

The most important factor influencing the maximum hardness that can be attained is the mass of the metal being quenched. In a small section, the heat is extracted quickly, thus exceeding the critical cooling rate of the specific steel. The critical cooling rate is the rate of cooling that must be exceeded to prevent formation of non-martensitic products (transformation products). As section size increases, it becomes increasingly difficult to extract the heat fast enough to exceed the critical cooling rate and thus avoid formation of transformation products. A typical condition is shown in Fig. 5.10, which illustrates the effect of section size on surface hardness and is a good example of the mass effect. For small sections up to 13 mm (0.5 in.), full hardness of about 63 to 65 HRC is attainable. As the diameter of the quenched piece is increased, cooling rates and hardness decrease because the critical cooling rate for this specific steel is not exceeded.

Induction Austenitization

Transformation to Austenite. The initial step in heat treating steel is austenization, the process of transforming the microstructure to austenite. Figure 5.3 shows that when a steel alloy is heated, the transformation to austenite under equilibrium conditions starts at Ac_1 and completes at the Ac_3 (sometimes referred to as the critical temperature). Equilibrium means that there is sufficient time at the specific temperature for transformation to complete. For example, when a 0.40% carbon steel is heated to 725 °C (1340 °F), its crystalline structure begins to transform to austenite.

If held for a long period of time at 725 °C (1340 °F) according to the phase diagram, the resulting microstructure will be a mixture of ferrite and austenite. No matter how long it is held at 725 °C (1340 °F), transformation to austenite is not complete until a temperature of approximately 815 °C (1500 °F) is reached.

In contrast, as shown in Fig. 5.3, steel containing 0.80% carbon transforms completely to austenite when heated to 725 °C (1340 °F) at what is called the eutectoid composition. The eutectoid composition is the only composition at which the Ac_1 and Ac_3 are at the same temperature. If the amount of carbon is increased to above 0.80%, such as to 1.0% carbon, and the steel is heated to 725 °C (1340 °F) or just above, austenite is formed. However, because only 0.80% carbon can be completely dissolved at this temperature, 0.20% carbon remains as cementite unless the temperature is increased to above 1040 °C (1450 °F). Increasing the austenization temperature will increase the amount of carbon that can be taken into solution, up to 2.00% carbon. The solution and diffusion of carbon are dependent on both time and temperature. Carbon is found in the cementite or carbide particles. Larger carbides take longer to dissolve, so that the practical effect on induction heating is that the fast heating cycles of induction heating may not dissolve all of the carbides. The difference between heat treatment temperatures for induction heating and furnace heating is specifically the inherent short austenitizing times when using induction. Three metallurgical factors affect the transformation temperatures for austenization with induction heating: (a) the chemical composition, (b) the prior microstructure, and (c) the rate of heating.

Effect of Chemical Composition on Transition Temperatures. The A_1 and A_3 transformation temperatures shown in Fig. 5.3 are for steel alloyed with carbon only. The addition of other elements has the effect of changing the effective phase diagram into a multi-dimensional diagram with movement of the A_1 and A_3. All important alloying elements decrease the eutectoid iron-carbon composition ratio and shift the A_1 and A_3. These shifts do not have much effect on induction austenitizing because workpieces must be heated to temperatures through and over the A_3. The prior microstructure and heating rate have more effect.

Effect of Prior Microstructure. As described in Fig. 5.3, heating for transformation into austenite has so far involved equilibrium conditions. When the prior microstructure does not have a uniform distribution of phases or elements, complete austenization requires more time at temperature, or a higher austenitizing temperature. Induction normally uses a timed heat cycle that does not hold the workpiece at temperature. Workpieces are heated to a specific temperature and are then cooled. Large carbide particles and some elements, such as manganese or nickel, diffuse very slowly. Accordingly, due to lack of time for solution and diffusion, complete transformation into a homogeneous austenitic phase can be difficult with the fast austenitizing cycles of induction heating. Higher

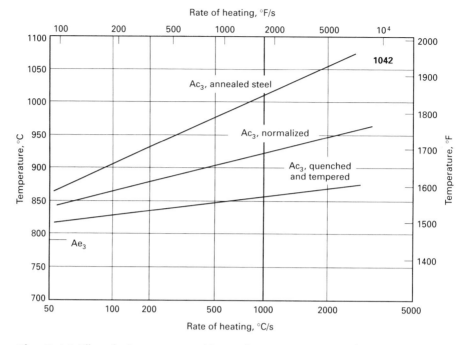

Fig. 5.11 Effect of prior structure on AC_3 transformation temperature of 1042 steel. AC_3, temperature at which transformation of ferrite to austenite is completed on heating. Source: Ref 4

austenitizing temperatures help because the rate of diffusion roughly doubles for every 6 °C (10 °F) that carbon steels are heated over the A_1. Therefore, higher austenitizing temperatures are used when induction heat treating than when furnace heat treating. Figure 5.11 shows the effect of both prior structure and rate of heating on Ac_3, the upper critical temperature, for complete transformation of 1042 steel. Note that induction austenitizing often requires temperatures 91 to 114 °C (150 to 200 °F) higher than recommended furnace austenitizing temperatures. Table 5.7

Table 5.7 Approximate induction austenitizing temperatures for carbon and alloy steels

Carbon content, %	Temperature for furnace heating		Temperature for induction heating	
	°C	°F	°C	°F
0.30	845–870	1550–1600	900–925	1650–1700
0.35	830–855	1525–1575	900	1650
0.40	830–855	1525–1575	870–900	1600–1650
0.45	800–845	1475–1550	870–900	1600–1650
0.50	800–845	1475–1500	870	1600
0.60	800–845	1475–1500	845–870	1550–1600
>0.60	790–820	1450–1510	815–845	1500–1550

Recommended austenitizing temperatures for a specific application will depend on heating rates and prior microstructure. Free-machining and alloy grades are readily induction hardened. Alloy steels containing carbide-forming elements (for example, niobium, titanium, vanadium, chromium, molybdenum, and tungsten) should be austenitized at temperatures at least 55 to 100 °C (100 to 180 °F) higher than those indicated. Source: Ref 4

shows examples of approximate induction austenitizing temperatures for carbon and alloy steels.

Austenitizing of Various Steels. The chemical composition and the prior microstructure affect required austenitizing temperature for various steel alloys. The influence on particular grades varies considerably and affects induction heat treating practice.

Carbon Steel. The development of induction heating pioneered the substitution of case hardened carbon steels for through-hardened alloy steels. One of the oldest rules in steel selection is to select the grade of steel in which carbon content is chosen to produce the hardness and physical properties that are needed. The use of lower carbon steels maximizes ductility while reducing quench-cracking tendencies. In principle, if a hardness range of only 40 HRC is needed, the carbon content can be below 0.30 carbon. However, if the hardness needs to be greater than 60 HRC, a carbon content greater than 0.50 is preferred. From a machinability viewpoint, higher carbon in steel increases the difficulty of machining. For this reason, there is considerable use of resulphurized steels, such as 1141 and 1144, in which the sulphide inclusions produce a remarkable increase in machinability. The austenitizing temperature selected needs to be high enough for complete solution of the carbon and complete diffusion. Segregation will increase the required austenitizing temperature. Grain growth is not usually a problem unless the steel is substantially overheated.

Alloy steels require a prior microstructure that is quenched and tempered in order to quench to optimum hardness and to produce the case depths that are projected by the hardenability curves. Overheating produces a higher tendency toward quench cracks.

Grey Cast Iron. Because irons can vary significantly in total carbon content, a minimum combined carbon content of 0.40 to 0.50% (as pearlite) is recommended. During austenitizing, the pearlite will behave essentially the same as the carbon in steel of similar carbon content. Care needs to be taken when austenitizing because the melting point of grey iron is much lower than that of steel. A phosphide constituent (steadite) melts at 954° to 982 °C (1750 to 1800 °F). The melting is not easily visible during heating, and a melted part is usually nonconforming and not salvageable. While as-cast parts can be heat treated to lower hardnesses, some shapes and parts will crack during heating unless stress relieved prior to induction hardening.

Ductile Cast Iron. The response of nodular iron is dependent upon the amount of pearlite in the matrix. The recommended microstructure is normalized and tempered with the hardness from 225 to 280 BHN.

Stainless steels and tool steels usually require up to 110 °C (200 °F) higher than the furnace austenitizing temperatures, and they generally quench 1 to 2 points HRC lower than when furnace austenitizing. Overheating produces retained austenite with resulting lower as-quenched hardness. Parts with retained austenite must be cryogenically processed as soon as possible for reduction of the retained austenite. Because of the combination of high austenitizing temperature that is required throughout the entire case, deep case depths with full transformation are difficult to produce without overheating the surface temperature in order to austenitize deeper into the case.

The Transformation of Austenite

Thus far, the discussion has been confined to heating of steel and the effects of various alloys, heating times, and austenitizing temperatures. Following austenization, steel can be hardened to various degrees by the control of cooling rate and temperature at which the steel is held below the A_1 temperature for a given period of time before further cooling. The transformation of steel from austenite when cooled is the next area of discussion.

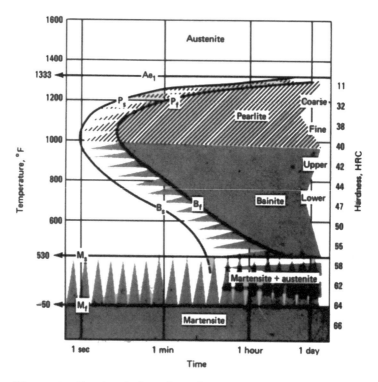

Fig. 5.12 Complete isothermal transformation diagram for 0.80% carbon steel. All of the transformation products are named. Bainite transformation takes place isothermally between 275 and about 525 °C (530 and about 975 °F). If austenite is rapidly cooled from above A_1 past the nose of the curve and to temperatures below 275 °C (530 °F), M_s, martensite starts to form. As long as cooling continues, more martensite forms. Transformation of austenite to martensite is not complete until M_t is reached. Ae_1, equilibrium transformation temperature in steel. Source: Ref 2

The temperatures at which the transformations occur on cooling are determined for conditions of slow cooling and for given carbon contents. It must be remembered that there are only three phases in steels, but there are many different structures. The eutectoid composition is the only composition to which carbon steels, on slow cooling, transform simultaneously into one constituent, pearlite. For the objectives of this book the precise amount of carbon denoted as eutectoid is of no particular significance and will be considered to be 0.80%. To better understand the transformation of austenite on cooling, these three definitions of carbon steel according to carbon content will be used: under 0.80%, 0.80%, and above 0.80%. The hardening reaction and the resulting microstructures, or transformation products, are best understood by studying the transformation theory as related to the effect of cooling rapidly and holding at temperature, and from continuous cooling, and varying the cooling rate. The isothermal transformation (IT) diagrams show the effect of equilibrium conditions on cooling, while the continuous cooling (CT) diagrams show the effects of continuous cooling.

The Isothermal Cooling Diagram, otherwise known as the Time-Temperature-Transformation Diagram (TTT), (Fig. 5.12) uses thin slices of the steel alloy which are cut and austenitized. The slices are then very rapidly quenched and are held at the quenched temperature for fixed time periods, with the time periods increased until transformation is complete. At the end of the fixed time periods the slices are rapidly quenched below the M_t temperature so that any remaining austenitic microstructure is transformed to martensite. The final microstructure will consist of whatever microstructure was produced during the time period during which the slice was held at higher temperature, plus martensite. Many test specimens need to be processed in order for the results of all temperatures and times held at temperature to be tested. Note that transformation at temperatures just under the Ae_1 require extensive time, while at 540 °C (1000 °F) only one second is needed. Pearlite is produced when quenched and held above 540 °C (1000 °F), while bainite is produced when held at lower temperatures. Quenching fast enough to miss the knee of what is termed the *C-curve*, where transformation starts, produces the desired martensitic microstructure. Notice also that according to equilibrium conditions, the test parts need to be quenched and to be held beneath the M_f temperature in order to have 100% martensite on final cooling. The importance of this diagram is that it shows the relative temperatures at which the different transformation products are produced and the importance of quenching fast enough to miss the "knee" of the diagram.

Transformation of Austenite for Steels under 0.80% Carbon (hypoeutectoid steels). In Fig. 5.3, the area shown on the diagram that is bounded by *AGB* is of significance to the room temperature; that is, microstructures of the steels with carbon contents under 0.80%. Within

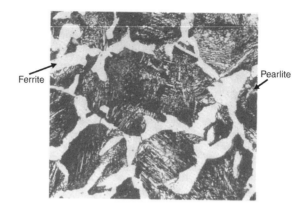

Fig. 5.13 Microstructure of pearlite and ferrite. 0.55% carbon steel slow cooled to form ferrite and pearlite. Picral etch, 500×. Source: Ref 7

this area the two phases of ferrite and austenite, each having different carbon contents, can exist at the same time.

Assume that a 0.40% carbon steel has been slowly heated until its temperature throughout the piece is 870 °C (1600 °F), thereby ensuring a fully austenitic structure. Upon slow cooling, free ferrite begins to form from the austenite when the temperature drops across the line *AG* into the area *AGB*, with increasing amounts of ferrite forming as the temperature continues to decline while in this area. Ideally, under very slow cooling conditions, all of the free ferrite separates from austenite by the time the temperature of the steel reaches A_1 (the line *BG*) at 725 °C (1340 °F). The austenite islands, which remain at about 725 °C (1340 °F), now have the same amount of carbon as the eutectoid steel, or about 0.80%. At or slightly below 725 °C (1340 °F) the remaining austenite transforms to pearlite. Upon cooling to room temperature there are no further changes in the crystalline structure. The final microstructure is a mixture of the ferrite and pearlite for a 0.56% carbon steel, as shown in Fig. 5.13. The ferrite does not etch and remains white, while the iron-carbide plates in the pearlite etch black.

Transformation of Austenite for 0.80% Carbon Steels (Eutectoid). 0.80% carbon steels transform to austenite at one transformation temperature, 725 °C (1340 °F) as shown in Fig. 5.3. The eutectoid temperature represents the lowest temperature of any carbon content in which complete transformation to austenite can occur. When the austenitic solid solution is slowly cooled, several changes occur at 725 °C (1340 °F) (the transformation temperature or critical temperature of the iron-cementite system). At this temperature, a 0.80% carbon steel transforms from a single homogeneous solid solution simultaneously into two distinct new solid phases of ferrite and cementite that are called pearlite. The change occurs at constant temperature and with the evolution of heat. This simultaneous formation of ferrite and cementite occurs only at the eutectoid composition point *G* in Fig. 5.3.

Fig. 5.14 Pearlite. 0.55 carbon steel austenitized at 860 °C (1580 °F), transformed at 650 °C (1202 °F) for 70 s. 300 HV. Picral. 250×. Source: Ref 7

The microstructure of a typical eutectoid steel is shown in Fig. 5.14. This represents a polished and slightly etched specimen at a magnification of 250X. The white matrix is alpha ferrite, and the dark platelets are cementite. All grains are pearlite—no free ferrite grains are present under these conditions. Cooling rates and temperatures govern the final condition of the particles of cementite that precipitate from the austenite at 725 °C (1340 °F). With normal cooling, the microstructure will be entirely pearlitic. Under specific cooling conditions under which the cooling is sufficiently slow or interrupted, the cementite particles become spherical instead of elongated platelets, as shown in Fig. 5.15. The microstructure appears as a matrix of ferrite with cementite or iron carbide globules.

Transformation of Steels from 0.80 to 2.0% Carbon (Hypereutectoid Steels). Steels of more than 0.80% carbon transform differently under equilibrium conditions. Assume that a steel containing 1.0% carbon has been heated to 845 °C (1550 °F), thereby ensuring a 100% austenitic

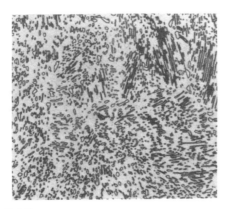

Fig. 5.15 Spheroidizing. Picral etch. 1000×. Source: Ref 7

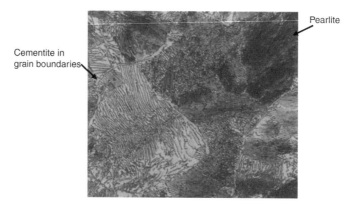

Fig. 5.16 Cementite network around pearlite. Proeutectoid cementite and pearlite formation. Picral etch. 500×. Source: Ref 7, p 268

microstructure. When cooled, no change occurs until the line *GF (Fig. 5.3)*, known as the A_{cm}, is reached. At this point cementite begins to separate out as the temperature of the steel descends below the *A* line. The composition of austenite changes from 1.0% carbon down towards 0.80% carbon as the carbon content is depleted by the formation of cementite. At a temperature slightly below 725 °C (1300 °F), the remaining austenite will all transform to pearlite because it is then at the eutectoid composition. No further changes occur as cooling proceeds toward room temperature, so that the room temperature microstructure consists of pearlite and free cementite. In this case the free cementite exists as a network around the pearlite grains (Fig. 5.16).

Fast Cooling of Austenite

The previous transformation products, as described, were produced under very slow cooling or equilibrium conditions. Induction heat treating practice uses rapid cooling (quenching) from the austenitic condition. With fast cooling, workpieces can be quenched to the preferred martensitic microstructure. If the cooling is not sufficiently fast, the final microstructures may have pearlite or bainite in the martensite. The effect of cooling rates and their effect on the transformation of austenite and formation of the various transformation products will be described next.

Continuous Cooling Transformation (CCT) Diagram. As mentioned, the IT diagram shows the microstructures produced by cooling and holding under equilibrium conditions. The diagrams have typically a C-shape in the curve, with time for transformation decreasing below the A_1, then increasing again below the C of the curve. With induction hardening, the cooling or quenching of the alloy needs to be fast enough that the knee of this C-curve is missed. The IT diagrams can illustrate the need for quenching fast enough to miss the "knee" of the C-curve, but they do not provide the dynamic model that is needed. Continuous Cooling (CCT) Diagrams show the microstructures produced by various rates of cooling. They help in selection of quenchants with fast enough quenching speeds to produce the cooling rates needed for martensitic microstructures. The CCT in Fig. 5.17 shows the continuous cooling of two different section sizes of 4340 steel, with the surface and center rates of cooling measured, and the final microstructure as-quenched. Note that the cooling rate of both the surface and center of the smaller diameter bar is fast enough to produce martensite; however, the center of the larger bar did not cool fast enough, and some bainite was produced. The larger section size quenched more slowly in the core, thus producing a slower cooling rate. Each steel alloy has its individual IT and CCT diagram, with the location and size of the C-curve depending upon the alloy content. Higher alloyed steels, completely austenitized, do not need to be cooled as fast as the carbon steels to quench to martensite. The theory and importance of quenching rate is discussed in the Chapter "Quenching" in this book.

Microstructures Produced during Cooling. The products formed during cooling from austenite are called transformation products. The products that are formed depend upon the steel alloy used and the cooling rate. Martensite is the microstructure normally desired for optimum mechanical properties. The other transformation products are considered undesirable. Transformation products discussed in the order of their potential formation when cooled from austenite, are:

- **Pearlite** is normally produced during slow cooling. The mechanism of formation is as described previously in the transformation of steels under 0.80% carbon. Pearlite is a mixture of the two phases of ferrite and cementite, with the cementite formed as lamellar plates in a matrix of ferrite. Pearlite can be found in the microstructure of a quenched part if the austenization was not complete. Most quenching rates are fast enough to avoid pearlite formation. The experience of the author is that pearlite is not normally found in induction-hardened parts unless there is interrupted or slack quenching or unless diffusion of the pearlite was not completed during austenization.
- **Bainite** can be produced in two forms: upper bainite and lower bainite. At temperatures lower than 565 °C (1050 °F) the urge for the austenite to transform is overcome by the lack of mobility of the carbon atoms. The austenite would like to transform, but the carbon atoms cannot move fast enough. Because the carbon atoms cannot move freely, the time necessary for transformation to begin is longer

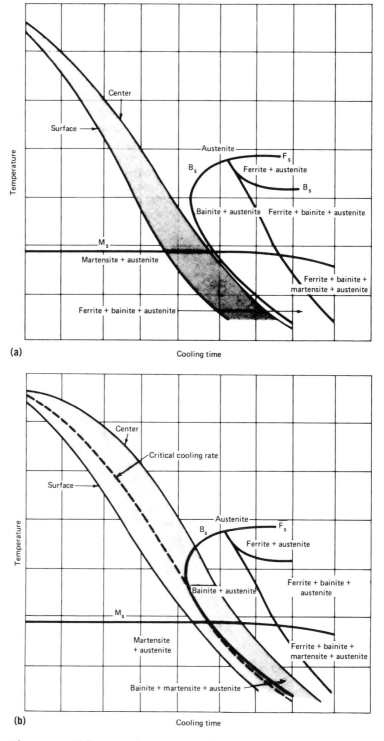

Fig. 5.17 IT diagram and cooling curve for 4340 steel. Effect of section size. (a) The cooling curve for a 25.4-mm (1-in.) diam AISI 4340 steel bar is shown in relation to the CCT curve for AISI 4340 steel. The cooling rate is sufficiently rapid to quench the entire bar to martensite. (b) A 76-mm (3-in.) diam bar is shown in relation to the same CCT curve. It cools more slowly under identical quenching conditions, and because of the mass, the center consequently transforms partially to ferrite and bainite. Source: Ref 2

as the temperatures fall lower than 565 °C (1050 F). Figure 5.12 shows the region for bainite formation for a 0.80% (eutectoid) carbon steel to be between 525 to 275 °C (975 to 530 °F). Bainite is similar to pearlite in that it has a lamellar mixture of ferrite and cementite, but the mixture is much finer. High magnification is often needed to accurately define bainite microstructures, as opposed to pearlitic microstructures. Upper bainite, formed at higher temperatures, has a feathery appearance. Lower bainite, formed at temperatures under upper bainite, has an acicular appearance. Figure 5.18 shows a microphotograph of lower bainite, which has the appearance of flakes. Because many manufacturers do not want bainite in induction-hardened parts, further discussion of bainite is given in Chapter 14.

- **Martensite** is a transformation product formed spontaneously by the rapid cooling of a steel alloy past approximately 275 °C (530 °F), the M_s temperature, to form a phase that is essentially a supersaturated solution of carbon in ferrite. A martensite crystal is formed by a mechanism that can be thought of in terms of shear that moves iron atoms co-operatively and almost simultaneously from their original sites in the austenite structure to the nearest available site in the final structure. Martensite plates have a needle-like or acicular outline when viewed by a microscope. The amount of transformation is essentially independent of time. On the IT and CT diagrams martensite formation starts at the M_s temperature and completes at the M_f temperature. Transformation to martensite is accompanied by a marked volume increase. It is difficult, if not impossible, to determine the M_f temperature with any accuracy because it is hard to distinguish small volumes of untransformed austenite by light microscopy. The end of the transformation is, therefore, often arbitrarily defined by either 90% or 95% transformation, which in the case of eutectoid steels happens to be close to room temperature. The M_f temperature of some steels is below room temperature. In a

Fig. 5.18 Lower bainite. Picral etch. 250×. Source: Ref 7

quenched eutectoid steel some 5 to 10 austenite volume percent is usually retained at room temperature. As-quenched and etched, the appearance is that of a white matrix that will have only oxide or sulfide inclusions showing. Figure 5.19(a) and (b) show an induction-hardened 1052 steel wheel-hub that quenched to 62.1 HRC. The as-quenched martensite at 200× (Fig. 5.19a) has little definition, while at 1000× (Fig. 5.19b), the acicular structure is prominent, and the clusters show some evidence of the original austenitic grain size plus a small titanium carbide inclusion just off center, which is actually slightly golden in color. The Chapter "Nonconforming Product and Process Problems" in this book discusses etching and inspection methods for martensitic microstructures.

- **Tempered Martensite.** The recommended heat treating practice for optimum physical properties is to quench a workpiece to a fully martensitic microstructure and then, if the hardness is higher than specified, to temper to reduce the hardness into the specified range. The Chapter "Tempering" in this book discusses tempering theory. Upon reheating a workpiece above 121 °C (250 °F), the martensite is not stable, and the carbon in solution in the martensite starts to precipitate. This microstructure after tempering is called *tempered martensite*, and it etches to a darker color with increasing tempering temperature. Figure 5.20(a) and (b) shows a specimen of the same steel that was shown in Fig. 5.19 that was tempered at 206 °C (400 °F), reducing the hardness to 62 HRC.
- **Retained Austenite.** For most steels above 0.60% carbon the M_f temperature, which is the temperature at which formation to martensite is complete, is below room temperature. Therefore, for these alloys a small percentage of residual, or retained, austenite is found when the alloys are quenched to room temperature. Small amounts of retained austenite are impossible to see by microscopy. Methods for reduction of retained austenite are discussed in the Chapter "Quality Control" in this book in the discussion of materials and process problems that are involved when the amount of retained austenite after quenching is objectionable.

Residual Stress and Induction Heat Treating

Residual stresses are stresses that exist in a solid body without an imposed external force. Factors such as cold working, phase changes while heat treating, and temperature gradients that occur during heating or cooling produce the stresses. The nature of stresses in a workpiece before induction hardening can influence distortion and cracking tendencies during heating and quenching. Often individuals question what can cause these effects. The answer is a lot. Table 5.8 shows a summary of the

(a)

(b)

Fig. 5.19 Untempered martensite. (a) 1052 steel induction hardened with 14% polymer quench, nital etched, 200×. The as-quenched hardness is HRC 62.1. (b) The same part at 1000×. Source: Courtesy of Aston Metallurgical Services, Inc.

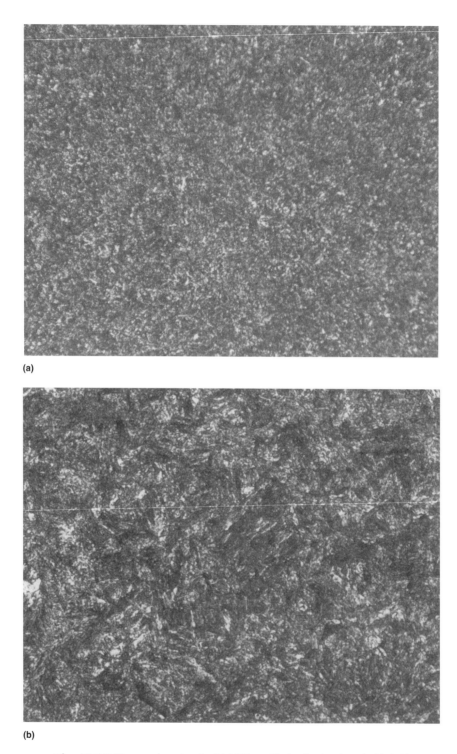

Fig. 5.20 Tempered martensite. (a) 1052 steel induction hardened with 14% polymer quench and tempered at 206 °C (400 °F), nital etched, 200×. The as-quenched hardness is HRC 56.2. (b) The same part at 1000×. Source: Courtesy of Aston Metallurgical Services, Inc.

Table 5.8 Summary of compressive and tensile residual stresses at the surface of the parts created by the common manufacturing processes

Compression at the surface	Tension at the surface
Surface working: shot peening, surface rolling, lapping, and so on	Rod or wire drawing with deep penetration
Rod or wire drawing with shallow penetration(a)	Rolling with deep penetration
Rolling with shallow penetration(a)	Swaging with deep penetration
Swaging with shallow penetration(a)	Tube sinking of the outer surface
Tube sinking of the inner surface	Plastic bending of the shortened side
Coining around holes	Grinding: normal practice and abusive conditions
Plastic bending of the stretched side	Direct-hardening steel (through-hardened)(b)
Grinding under gentle conditions	Decarburization of steel surface
Hammer peening	Weldment (last portion to reach room temperature)
Quenching without phase transformation	Machining: turning, milling
Direct-hardening steel (not through-hardened)	Built-up surface of shaft
Case-hardening steel	Electrical discharge machining
Induction and flame hardening	Flame cutting
Prestressing	
Ion exchange	

(a) Shallow penetration refers to ≤ 1% reduction in area or thickness; deep penetration refers to ≥ 1%. (b) Depends on the efficiency of quenching medium. Source: Ref 4

effects of compressive and tensile residual stresses at the surface of the parts created by common manufacturing processes.

There are two basic types of residual stresses: macrostresses, which are an average of body stresses over all large regions as compared to grain size; and microstresses, which are the average stress across one grain, or part of the grain, of the material. These two types of residual stresses may be classified further as tensile or compressive stress, with their locations starting at the surface and moving toward the center of a part.

Heat treatment can result in the development of residual stresses (both compressive and tensile), dimensional change (with respect to size and shape), and quench cracking. The major benefit of reduction of the overall residual stresses in a part is the elimination of dimensional change during manufacturing. The major benefit of producing compressive residual stress through induction hardening includes improved fatigue life and increased resistance to crack initiation. Tensile residual stresses at the surface of a part are generally undesirable because they can effectively increase the stress levels, as well as cause fatigue failure, quench cracks, grinding checks, and reduce the strength of a part. Excessive tensile residual stresses in the interior of a component may also be damaging because of the existence and consequence of defects that serve as stress raisers. Subsurface quench cracks can initiate due to excessive tensile stresses.

Measurement of residual stresses is generally done after a part is manufactured. There are two broad methods for the measurement: mechanical and physical. The mechanical methods include the stress relaxation methods of layer removal, cutting, hole drilling, and trepanning that incorporate the use of strain gauges and measurement of the movement of a part as the stressed layers are removed. The physical methods include x-ray diffraction, neutron diffraction, acoustic, and magnetic, with x-ray diffraction probably being the most widely used.

Processes that Influence Dimensional Change and Residual-Stress Patterns. A number of factors affect the residual stresses found in a completed part. Broad classifications are:

- Forming or metalworking processing before heat treating
- Heat treating operations before induction heat treating
- Effect of temperature gradients during heating and cooling
- Phase transformation during cooling
- Post quench tempering

Residual Stresses in Cold-Drawn Steel. The process of cold drawing steel produces mechanical stresses that are dependent on the shape of the dies and draft reduction as well as the prior microstructure, hardness, and grade of steel. Figure 5.21 shows the distribution of stresses in 1045 steel reduced by 20%. Also shown are the effects of both rotary straightening and stress relieving after cold drawing because there are significant effects on the residual stress pattern of the resulting products. The figures indicate the center of the bar and the stress levels toward the surface. Negative levels are compression, and positive levels are tension. Note the overall balance in a bar before machining is zero. All tensile levels balance with the compressive levels. However, once a part is machined, the levels are changed. The surface can be left in either compression or tension, depending on the previous stress condition and how much machining is done.

Residual Stresses Caused by Machining. Many complex parts are made from machining forgings, castings, bars, or plates. Stresses that are present from either castings or forgings that are machined without prior heat treatment can affect the final residual stress balance in a machined part. The presence of residual stresses in the workpiece affects its machinability; on the other hand, the machining process also creates and changes the residual stresses and produces undesired distortions in the part. Machine-shop personnel may have to experiment with a number of process parameters to minimize detrimental residual stresses such as the prior heat treatment, type of cutting tool, depth of cut, speed of turning, coolant, clamping/unclamping techniques, machining sequence, and even the frequency of tool sharpening or change.

Stress Relief. The basic objective of stress relieving is to produce the rearrangement of atoms or molecules from their momentary equilibrium position to more stable positions associated with lower potential energy

Heat Treating Basics / 111

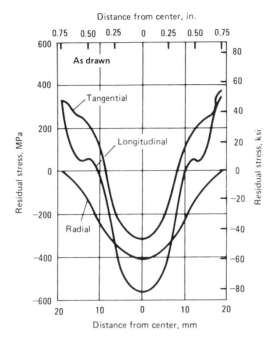

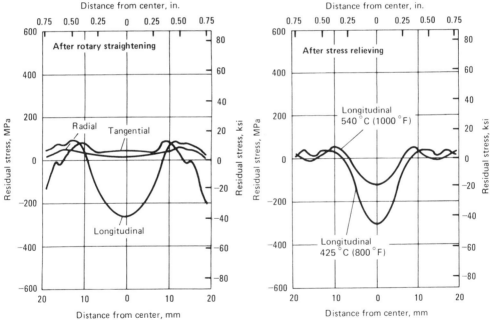

Fig. 5.21 Residual stress patterns in cold drawn steel bars of 1045 steel. Bars were cold drawn 20% from 43 to 38 mm (1¹¹⁄₁₆ to 1½ in.). Source: Ref 5

or stress state. Stress relieving operations for parts that are to be induction hardened involve heating the part to a certain temperature and holding at the elevated temperature for a specified length of time, followed by cooling to room temperature. Stress relief of parts to relieve cold-worked stresses is usually done in the 370 to 649 °C (700 to 1200 °F) range. Low-temperature tempering of quenched steels is sometimes done at 120 to 150 °C (250 to 300 °F) and is called or termed stress relieving because ductility in increased. If parts are machined after quenching and tempering, the stress relieving must not be done at a temperature higher than the original tempering condition, or there will be a decrease in hardness.

Stress Relieving of Cold-Drawn Bars. The process of heating for stress relieving (thermal processing) of cold-drawn bars is also known as *strain drawing, strain tempering, strain annealing, strain relieving, pre-aging,* or *stabilizing*. Thermal stress relieving is probably the most widely used treatment applied to cold drawn bars. Its purpose is to modify the magnitude and distribution of residual stresses in the cold-finished bar and thereby to produce a product with the desired combination of mechanical properties for field service. When at the stress relieving temperature, the elastic-strain energy is released in small but significant amounts of plastic deformation. Temperatures up to 650 °C (1200 °F) are used, with most of the commercial temperatures ranging from 370 to 480 °C (700 to 900 °F). When stress relieving is performed at relatively low temperatures, hardness, tensile strength, and elastic properties of most cold-drawn steels are increased. At higher temperatures, however, hardness, tensile strength, and yield strength are reduced. The choice of a specific time and temperature is dependent on chemical composition, cold drawing practice, and the final properties required in the bar.

Residual Stresses due to Heating and Cooling. Thermal stresses are produced first during the heating and then during the cooling of a part. Figure 5.22 shows the stresses at the surface of a carbon steel cylinder during heating and quenching. During heating, the surface remains in compression until the temperature reaches 1000 °C (1830 °F) when the surface becomes plastic. During cooling, the surface stresses are in tension until the surface is in the martensite formation range. The level of stress produced depends on the modulus of elasticity and the thermal coefficient of expansion of the material. The very basic nature of induction heating involves the conduction of heat from the surface toward the core. The rate of heating and the amount of heat transfer determine the thermal stresses during heating, as well as any stress relief of pre-existing stresses in the core. Residual stresses on heating have little effect, except when tensile stresses are high on the inside of a part, and the outside compressive

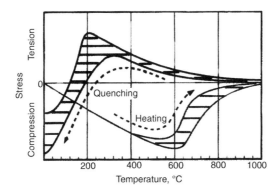

Fig. 5.22 Stresses at the surface of a carbon steel cylinder during heating and quenching. Martensite formation results in a final compressive residual stress state. Source: Ref 8

stresses are relaxed through subsequent plastic deformation so that the inside remaining tensile stresses are large enough to cause the part to crack. Induction heating can cause cracks when heating cold-drawn bar stock, when the bar is not stress relieved uniformly and there are still large tensile stresses inside the core.

During the cooling of induction-hardened parts, different stress conditions are set up. First, due to thermal contraction, the outside surface is increasing in tensile strength (the ductility is decreasing), while trying to decrease in volume. The resultant net residual stresses in the part after cooling depend on the contraction of the core and its effect on contraction of the surface. If the parts are case hardened fast enough so that the core has little heat, the core may have little effect on surface stresses. If the part is through heated so that the core has heat and can contract when cooled, there may be substantial effect. The final residual stress distribution with induction heating is more complex when the quenched parts undergo phase transformation to martensite.

Changes in Volume due to Phase Transformation. Figure 5.23 shows the compressive stresses in the case of an induction-hardened part. The most important change is the change of austenite into martensite, in that there is an approximate increase in volume of 4% times the carbon content. The area that is martensitic tries to increase in size, producing the compressive stress as the core cools. Surfaces can quench in tension when there is very little temperature variation between the surface and center of the part.

Residual stress patterns due to thermal and transformation volume changes are complex because the change in stresses due to thermal volume are independent of the changes caused by phase formation. During quench hardening of a steel workpiece, hard martensite forms at the surface layers, associated with thermal volume contraction and

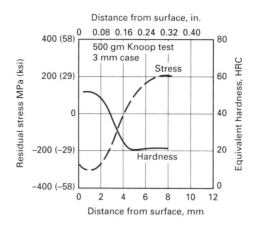

Fig. 5.23 Typical hardness and residual stress profile in induction-hardened (to 3 mm, or 0.12 in., case depth) and tempered (at 260 °C, or 500 °F) 1045 steel. Source: Ref 4

phase-volume increase. The remainder of the workpiece under the surface is still hot and ductile austenite (assuming the entire part was austenitized). Later, the remaining austenite transforms to martensite, buts its volumetric expansion is restricted by the hardened surface layer. This restriction causes the central portion to be under compression, with the outer surface under tension. During the final cooling of the interior core, surface contraction is hindered by the hardened surface layers. This restraint in contraction produces tensile stress in the core and compressive stress at the outer surface. Figure 5.24 illustrates the complex pattern of residual stress distribution over the diameter of a quenched bar, showing residual stresses after quenching in the longitudinal and radial directions in relationship to the coil and area heated. In some particular conditions, the volumetric changes can produce sufficiently large residual stresses that can cause plastic deformation on cooling, leading to warping or distortion of the steel part. Also, while plastic deformation appears to reduce the severity of quenching stresses, with severe quenching the quenching stresses are so high that they do not get sufficiently released. Consequently, the large remaining residual stresses can reach or even exceed fracture stress of steel and can cause quench cracks.

Residual Stresses in Induction Hardening. After quenching an induction-hardened part, compressive residual stresses may develop at the surface when martensite forms on the surface. Development of these compressive stresses depends on the nature of the hardened case and the stresses developed during quenching. The magnitude of the compressive stress effectively increases the yield strength, permitting the application of significantly higher stresses than could normally be possible in fatigue loading. It should be noted that there is a sharp transition to a tensile state

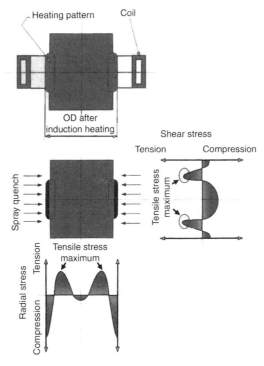

Fig. 5.24 Complex pattern of residual stresses forms in a carbon steel cylinder after induction heating and spray quenching. One of the goals of tempering is to relieve the subsurface tensile stresses that can cause cracking in service. Surface compressive stresses are beneficial. Note: Stresses shown are on a macroscopic scale. OD, outside diameter. Source: Ref 8

in the transition zone between the case and unhardened core material. In induction hardening, an increase in hardenability changes the depth at which transition from compressive to tensile stress can occur. In some applications an initial preheat to increase the core temperature, followed by increase in the rate of heating, can optimize and produce the maximum compressive stress at the surface while minimizing tensile stress under the case.

While the compressive residual stress on the surface is considered to be beneficial, the maximum tensile residual stress is located just beneath the hardened case. This is a zone of potential danger, because it is where most subsurface cracks initiate. The result is that the overall residual stress condition in the as-quenched part promotes brittleness that reduces the part reliability. Tempering is needed to relieve some of these stresses, ideally to remove or substantially reduce sub-surface tensile stresses, while retaining the beneficial surface compressive stresses. Finally, where compressive surface stresses are important, it should be noted that machining or grinding operations after induction hardening may need to be examined so that the compressive stresses are not removed with the part left in an essentially tensile mode.

Distortion

Distortion can be defined as an irreversible and usually unpredictable dimensional change in the part during heat treating. The term *dimensional change* is used to denote changes in both size and shape. For purposes of induction heat treatment, distortion is an uncontrolled, irreversible movement that occurs in a part as a result of the induction heat treating process. The intention of the induction heat treating is to keep the distortion within allowable limits that will enable the part to be completed as designed in the manufacturing process. However, sometimes processes such as tempering, cryogenic processing, straightening, machining, and grinding operations must be used to put the part back into useable tolerances. Figure 5.25 and 5.26 show two different types of parts with distortion problems. Both gears have distortion due to size change and shape change. Note that the recommendation for eliminating the distortion in Fig. 5.25 is to change the design of the gear, whereas for the gear in Fig. 5.26 the weight holes should be located deeper or should be drilled after the part is induction hardened.

Size Distortion. Change in size in the microconstituents due to the volumetric changes that occur during a phase change were previously discussed. Limiting size change can be very important when semi-finished parts are induction hardened and zero net change in size is an objective.

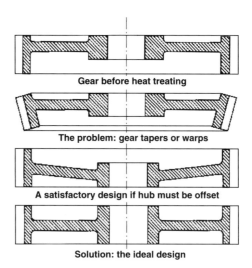

Fig. 5.25 Distortion caused by lack of symmetry in a gear. A typical problem caused by lack of symmetry in design, illustrated by a gear that warped during heat treating. Design modification can solve the problem. Source: Ref 9

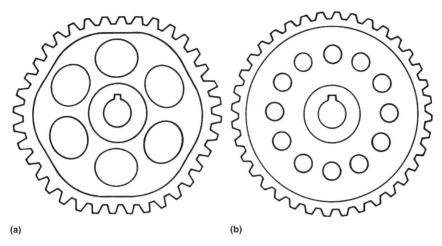

Fig. 5.26 Distortion caused by lack of symmetry in a gear. Problem caused by the use of holes to reduce weight of a gear. (a) If the designer specifies large holes in the web, heat treatment may produce a flat spot for each hole. (b) Keeping the hole diameter to one-third of the web width eliminates the problem. Source: Ref 9

Shape Change, Bending, or Warpage. The effects of the residual stresses in parts before they are induction hardened and the induction hardening process itself can cause distortion due to shape change. Spiral gears may straighten, and shafts may bend. Bores may close, and overall lengths may change. Potential methods for reduction of distortion will be discussed in the Chapter "Induction Heat Treating Process Analysis" in this book.

Grain Size

Grain size refers to the number of grains per unit area in terms of average diameter. Grain size can be important both before and after the induction hardening of a part. The grain size before hardening influences the hardenability of a material. The hardenability of a carbon steel may increase as much as 50% with an increase in austenite grain size from ASTM 8 (fine grained) to ASTM 3 (coarse grained). Grain-size requirements after induction hardening are important if there are specifications on the martensitic grain size after quenching. (In this case the martensitic grain size is the only way to measure the austenitic grain size before quenching.)

While there are a number of different procedures that can be employed to measure grain size, one of the more common methods is comparison with standard charts. The standard chart allows the parts to be viewed at 100× and compared to the charts to determine the grain size as illustrated in Fig. 5.27 and 5.28. Figure 5.29 shows microphotographs of four

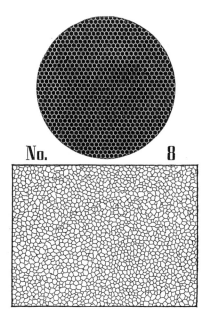

Fig. 5.27 Grain size No. 8. Upper, idealized hexagonal network for mean grain size No. 8, ASTM scale, 128 gr per sq. in. Lower, ASTM standard grain size No. 8, 96 to 192 gr per sq. in. at 50×. Source: Ref 10

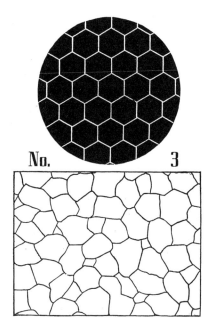

Fig. 5.28 Grain size No. 3. Upper, idealized hexagonal network for mean grain size No. 3, ASTM scale, 4 gr per sq. in. Lower, ASTM standard grain size No. 3, 3 to 6 gr per sq. in. at 50×. Source: Ref 10

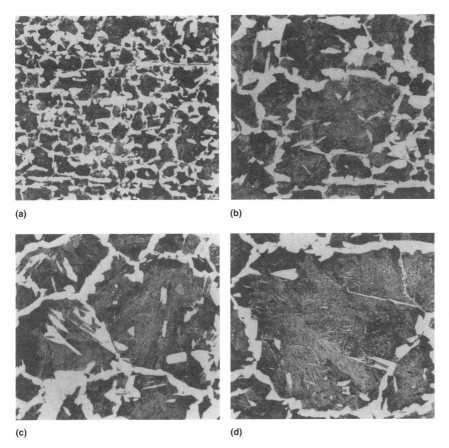

Fig. 5.29 Microphotos of grain sizes. Austenite grain growth in a normal 0.5% C hypoeutectoid steel (silicon deoxidized). 0.50C-0.06Si-0.7Mn (wt%). (a) austenitized for 1 h at 850 °C, cooled at 300 °C/h. Austenite grain size: ASTM: No. 5. 180 HV. Picral. 100×; (b) Austenitized for 1 h at 900 °C, cooled at 300 °C/h. Austenite grain size: ASTM No. 3. 180 HV. Picral. 100×. (c) Austenitized for 1 h at 1100 °C, cooled at 300 °/h. Austenite grain size: ASTM No. 1. 180 HV. Picral. 200×. (d) Austenitized for 1 h at 1200 °C, cooled at 300 °C/h. Austenite grain size: ASTM No. 0. 180 HV. Picral. 100×. Source: Ref 11

austenitic grain sizes ranging from grain size 5 to 0. It is often difficult to determine the prior austenitic grain size in a fully martensitic microstructure. When the austenitic grain size becomes large enough, sometimes the vague previous austenitic grain size can be seen through the orientation of different packets of martensite.

The parts are prepared metallurgically for grain size measurement. Preparation is discussed in the Chapter "Standards and Inspection" in this book.

REFERENCES

1. R.E. Haimbaugh, Induction Heat Treating Corp., personal research
2. Standard Practices and Procedures for Steel, *Heat Treaters Guide*, American Society for Metals, 1982

3. *Properties and Selection: Stainless Steels, Tool Materials, and Special Purpose Metals*, Vol 3, *Metals Handbook*, 9th ed., American Society for Metals, 1980
4. *Heat Treating*, Vol 4, *ASM Handbook*, ASM International, 1991
5. *Properties and Selection: Irons and Steels*, Vol 1, *Metals Handbook*, 9th ed., American Society for Metals, 1978
6. *Mechanical Testing and Evaluation*, Vol 8, *ASM Handbook*, ASM International, 2000
7. L.E. Samuels, *Light Microscopy of Carbon Steel*, ASM International, 1999
8. K. Weiss, Induction Tempering of Steel, *Adv. Mater. Process.*, Aug 1999
9. *Materials Selection and Design*, Vol 20, *ASM Handbook*, ASM International, 1997
10. G.L. Kehl, *Principles of Metallographic Laboratory Practice*, McGraw Hill, 1949
11. R.S. Rufini et al., Advanced Design of Induction Heat Treating Coils, *Ind. Heat.*, Nov 1998

CHAPTER **6**

Quenching

WHEN INDUCTION HARDENING, the workpieces are cooled by surrounding, spraying, or immersing the part in a quenchant, with the exception of a steel workpiece that has had its heat removed by air or self quenching. Quenching is the cooling of a workpiece at a controlled rate in order to obtain the desired microstructure and hardness. The Chapter "Heat Treating Basics" in this book covers the cooling of austenite and the subsequent transformation of austenite when cooled, with the microstructure produced during cooling dependent on cooling rate and steel composition. Hardness and hardenability are determined by the chemical composition of the steel, the microstructure when austenitized, and the cooling rate of the quenchant. Selection of the quenchant is governed by the cooling rate that is needed to quench the workpiece to full hardness without producing quench cracks, and which will minimize distortion (if distortion is a factor). This Chapter will discuss the theory and selection of quenchants.

Three Stages of Quenching

During the quenching of steel in liquid media, the process may be split into the following three stages: (a) the vapor blanket stage, (b) the boiling stage, and (c) the convection stage. Figure 6.1 gives examples of the three stages of quenching.

The vapor blanket stage occurs when the workpiece is first quenched for cooling from austenite. The very hot surface temperature of the workpiece vaporizes the quenchant, and a thin vapor pocket forms around the workpiece. Heat transfer occurs by radiation and conduction through the vapor blanket, which acts essentially as an insulating layer because of the relatively poor conductivity of a vapor. The vapor stage is a period of relatively slow cooling, and the cooling effect of various quenchants varies greatly due to the differences in the thickness of vapor pockets that are formed. Vapor pocket formation tendency is reduced through the use of cold water and through the use of spray quenches. Another way is to use additives to the water, such as salt and polymers. Finally, oil quenches produce virtually no vapor pocket. The initial heat extraction of oil quenches is fast, but the quenching speed slows substantially in the other stages of quenching.

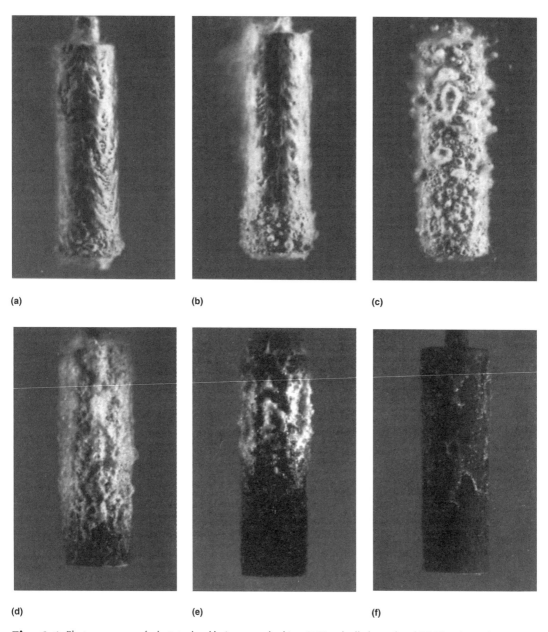

Fig. 6.1 Photo sequence of a hot steel rod being quenched in a 25% polyalkylene glycol (PAG) polymer in water solution. (a) When the rod is immersed, a polymer film forms on its surface. (b) After 15 s, polymer activates and begins to boil. (c) After 25 s, boiling occurs over the entire rod as the cooling rate increases. (d) After 35 s, boiling collapses and the convection stage begins. (e) After 60 s, the polymer starts to redissolve. (f) After 75 s, polymer film has completely redissolved, and the heat removal is achieved entirely by convection. Source: Ref 1

The boiling stage is the second stage of cooling. The vapor pocket collapses, and the quenchant comes into contact with the hot metal surface of the part, resulting in nucleate boiling to cool the part. This stage of quenching typically has the highest rate of heat extraction.

The convection stage is the third stage of cooling, and it starts when the surface temperature of the part being cooled decreases below the boiling point of the quenchant. Heat transfer occurs by direct contact between the surface of the part being cooled and by convection of heat through the quenchant. The cooling rate is low because of the low temperature differential between the part being cooled and the quenchant. The cooling rate is affected primarily by the quenchant viscosity, agitation, and convection rate.

Types of Quenchants

In the early years of induction heat treating, water was widely used for quenching, with salt additives used when severe quenchants were needed. In the 1950s the introduction of fast-speed oil quenchants permitted the use of oil as a quenchant for medium-carbon steels where there were cracking tendencies because of small section size. Later, various water soluble oils and plastic quenchants were introduced, followed by the introduction and use of the polymer-based water quenchants that continue to be in heavy use today. Polymer quenchants have eliminated spotty hardnesses that were caused by vapor pocket formation. Polymers also provide more flexibility when used for quenching induction austenitized workpieces.

The various quenchants used have different cooling characteristics in cooling rates. Different materials and part geometry have different quenching rate requirements in order to produce a fully martensitic microstructure. The quench severity of various quenchants versus cooling rates needed for different steel alloys is shown in Fig. 6.2. Low-carbon steels need high cooling rates, while the high-alloy steels can have slow cooling rates. (Workpieces made from 4340 steel will quench to high hardness in air.) The characteristics of the various quenchants follow:

Water has continued to be an excellent quenchant when used appropriately. Water is cheap and readily available. Figure 6.3 shows the cooling rate versus temperature for water quenchants with moderate agitation of the water. Note the substantial drop in cooling rate from 26 to 40 °C (80 to 100 °F). Figure 6.4 shows the effect of cooling rates of water at different temperatures throughout decreasing workpiece temperature. The fastest cooling rates are produced with the coldest water in the boiling stage. The quenching speed of water is much faster at or below 26 °C (80 °F). However, in practice where the quench water is recirculated through a water-to-water heat exchanger, the recirculated water temperature is more likely to be above the 32 °C (90 °F) range, unless the line cooling water is

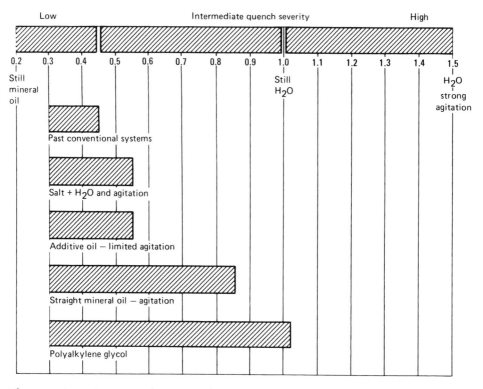

Fig. 6.2 Approximate quench severities for quenching mediums containing additives to improve cooling capacity. Source: Ref 2

obtained from a source such as a well or through a chiller. Even 6 °C (10 °F) can make a big difference in the quenching ability of water. The Appendix "Quench System Design" in this book discusses system design requirements for keeping quenchant temperatures in a controlled range in order to maintain the same cooling rate throughout the workday.

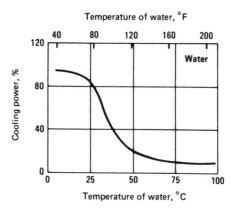

Fig. 6.3 Relation of surface cooling power of water with moderate agitation and water temperature. Source: Ref 3

Brine-based quenching systems have been used in place of water for the very high quenching rate requirements. Brine is made through the addition of salts to water quenching systems. Commonly used materials include sodium chloride (NaCl), typically at 10%, and sodium hydroxide (NaOH), typically at 3%. When a workpiece is first quenched, salt crystals form and deposit on the surface of the workpiece as shown in Fig. 6.5. Then the crystals fragment and cause disruption of the vapor film. Figure 6.6 shows an example of the effect of salt and the quenching characteristics of various salt solutions compared to water. The use of brine is limited because of high maintenance required for brine quenching systems, along with corrosion problems on the equipment and workpieces.

Oil. Through careful formulation and blending, a wide range of quenching characteristics can be obtained in oil quenches. Quenching oils for conventional heat treating are classified by three quenching speeds: normal-speed quenching oils, medium-speed quenching oils, and high-speed quenching oils. Oils have traditionally been used where there were distortion or cracking problems because of the slow cooling speeds of oil during the M_f-M_s transformation to martensite. As discussed in the Chapter "Heat Treating Basics" in this book, a workpiece is least likely to crack when the entire cross section being hardened transforms to martensite at the same time. Oil quench systems, as compared to the other quench systems, have a wide band of temperature range of operation in which the quenching speed does not vary enough

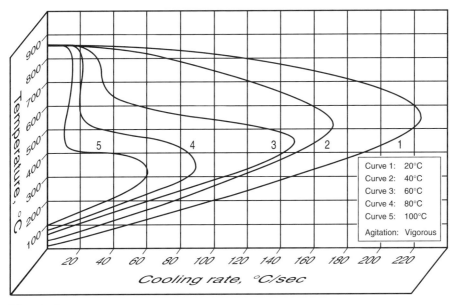

Fig. 6.4 The effect of temperature on the quenching characteristics of water. Source: Ref 4

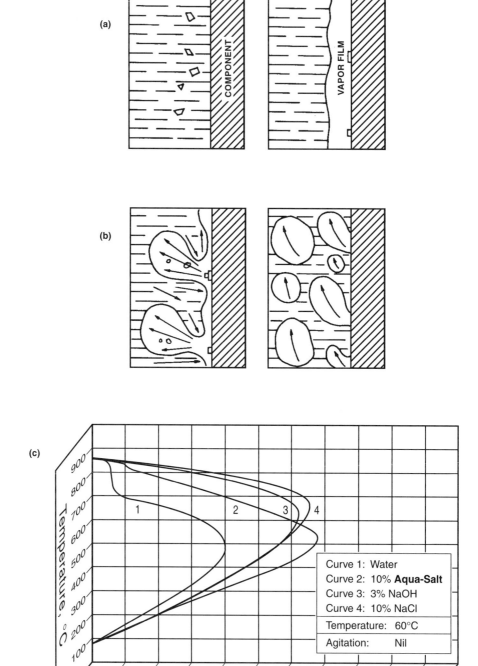

Fig. 6.5 (a) Salt crystals form and deposit on the component surface. (b) Crystals fragment, causing disruption of the vapor film. (c) The quenching characteristics of various salt solutions. Source: Ref 4

to affect the hardness of the workpiece). The viscosity of quench oil increases up to the 66 °C (150 °F) region, producing faster cooling. Quench bath temperatures can range without any problem from room temperature to over 100 °C (212 °F). When workpieces are immersed in oil, flames appear at the point of immersion. At that time the workpiece needs to be quickly, completely immersed in the oil bath so that the flame that was around the workpiece self extinguishes. The flash point of quench oils is as high as 182 °C (360 °F), so the quench bath will not ignite as long as the bath temperature is below this point. During the immersion of the workpiece, smoke is produced that may need to be exhausted. When used for spray quenching, however, the workpiece itself needs to be quenched below the flash point to prevent fires. After quenching with oil, the workpieces are normally cleaned to remove the oil dragout. Polymer-water based quenchants have taken the place of most oil quenches because of the flammable characteristics of oil and the smoke produced with spray quenches. Although some of the polymer quenchants claim to have the same cooling rate as oil in the martensitic transformation range, the curves are based on dunk quenching curves and testing is recommended when making a change from oil to polymer.

Polymer quenches consist of solutions of organic polymers in water. They also contain corrosion inhibitors and other additives to produce concentrates that are further diluted to give ready-to-use quenching solutions. Polymer quenches eliminate the danger of fire hazard that exists with oil.

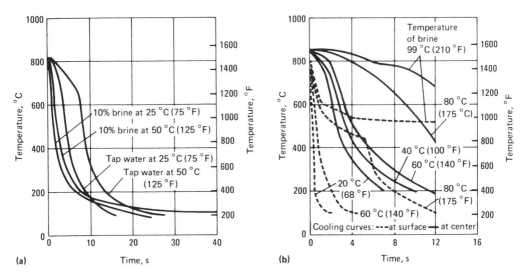

Fig. 6.6 Cooling curves showing effect of temperature on cooling power. (a) Center cooling curves for water and 10% NaCl solution in quenching 18-8 stainless steel specimens 13 mm (½ in.) diam by 64 mm (2½ in.). No agitation. (b) Center and surface cooling curves for 5% NaCl solution at 0.9 mm/s (3 ft/s) in quenching 0.95% C steel specimens 13 mm (½ in.) diam by 50 mm (2 in.). Source: Ref 3

At low concentrations the polymer quenches eliminate the soft spotting due to vapor pocket formation that is found with water quenches. (Note that proper quenching techniques must still be used, or self tempering of martensite can occur because of improper quenchant impingement.) One of the major advantages of polymer quenches is the ability to mix concentrations that cool anywhere from the quenching speed of water to the quenching speed of oil. Polymers can be selected, and concentrations can be mixed to produce the quenching speeds and characteristics needed for specific parts made of specific steels.

Types of polymers
used include several different types:

- PAG—polyalkylene glycol
- ACR—sodium polyacrylate
- PVP—polyvinyl pyrrolidone
- PEO—polyethyl oxazoline

PAG and PEO are the most commonly used polymers with induction.

Cooling curves furnished by the manufacturers for polymer quenchants cover concentrations typically from 0 to 30% in concentration. Normal polymer quenchant temperatures in induction heating systems are held within 6 °C (10 °F) of the bath temperatures between 32 and 43 °C (90 and 100 °F). The PAG and PEO polymers are the polymers used most commonly for induction. They have similar characteristics in that their quenching speed decreases with increasing concentration and decreases with an increase in quenching temperature. All polymers have an increase in cooling rate with increased agitation. PAG polymers have what is termed *inverse separation*. Depending on the chemical structure, the polymer that is first deposited on the steel part during cooling goes back into solution in the quenchant in the 60 to 90 °C (140 to 194 °F) region. Figures 6.7(a), (b), and (c) show the cooling rates of PAG quenchants as a function of concentration, temperature, and agitation. The PAG polymers have dragout, and with heavy concentrations not only do the workpieces feel "slimy," but also the fixturing in contact with the quenchant becomes coated.

The polymer-to-water ratio of PEOs is highly efficient in modifying quenching characteristics and required properties at low polymer concentrations. The inverse dragout is at 65 °C (150 °F), and PEOs have less dragout after quenching. Figures 6.8(a), (b), and (c) show the cooling rates of PEOs as a function of concentration, temperature, and agitation. PEOs have a faster cooling rate at low concentrations than the PAGs, while having a lower cooling rate in the martensite-formation temperature range. The PEO polymers are rated to have the most oil-like quenching characteristics of the polymer quenchants.

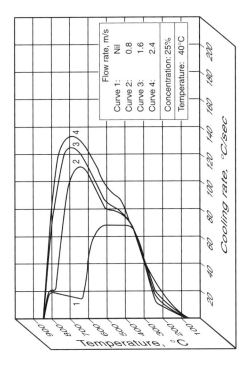
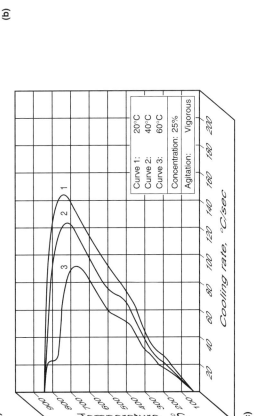

Fig. 6.7 Polyalkylene glycol quenchant (PAG) (a) The effect of concentration on quenching characteristics; (b) the effect of agitation on quenching characteristics; (c) the effect of temperature on quenching characteristics. Source: Ref 4

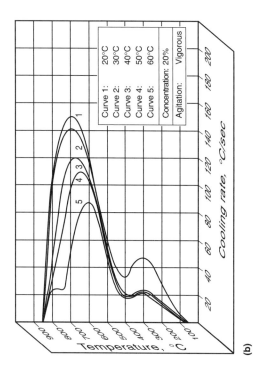
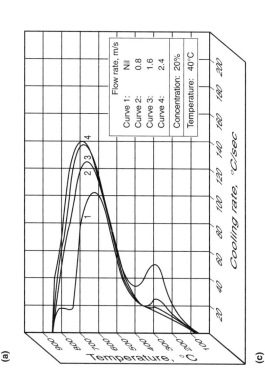

Fig. 6.8 Polyethyl oxazoline quenchant (PEO) (a) The effect of concentration on quenching characteristics; (b) the effect of temperature on quenching characteristics; (c) the effect of agitation on quenching characteristics. Source: Ref 4

Comparison of Quenchant Curves

One of the common modes for evaluating different quenchants is through the use of a quenchometer. Fixed diameter balls have thermocouples inserted so that the balls can be quenched with the cooling rate measured. The cooling curves produced from tests give comparisons among the speeds of different quenchants under different conditions as shown in the figures. The basis for initial selection of a quenchant for an induction-hardened part is provided by these curves. Figure 6.9 shows the comparison of the cooling curves of water, a 20% PAG polymer quench, and oil quench. These curves are all based upon dunk quenching of 25.4 mm (1 in.) diameter steel bars. Since most induction quenches are spray quenches, the effect of the direct impingement of the spray quench on larger workpieces may cause the quenchant to cool not only differently, but also faster—at least in the vapor cooling stage. Polymer quenchants have extensive use in polymer-to-water ratios ranging between 1.0% and 25% to produce quenching speeds between those of water and fast oil. As the polymer-to-water concentration is reduced, the faster cooling rate decreases the overall time of quenching, thereby increasing productivity. Testing is necessary with many workpieces so that the quench concentration is characterized to a controllable range in which the leanest ratio of quenchant can be used without fear of cracking.

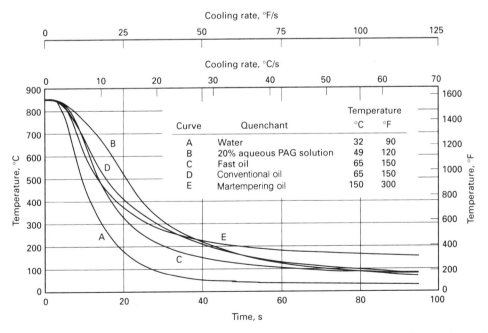

Fig. 6.9 Effect of selected quenchants on the cooling curve of a 25.4 mm (1.0 in.) diam steel bar. All quenchants flowing at 0.50 m/s (100 ft/min). Source: Ref 1

Quenching Methods

There are many methods of quenching used for induction heat treating. The easiest involves the use of either air or mass quenching. Some tool steels, such as D2, can be cooled in air to obtain full hardness. Other parts can have small areas quickly austenitized. When the heating cycle is complete, the cold mass under the part and adjacent to the heated area removes the heat quickly enough to cool the part to full martensitic transformation. As mentioned, this is usually done where the area can be very quickly heated, and the part is large enough to have enough mass and heat sink that the heated area is rapidly cooled.

Conventional quenching techniques used by furnace heat treating usually involve dunk quenching into a quench tank. Dunk quenching can be used in induction heat treating, particularly for alloy steels. Depending upon the material and workpiece size, the critical cooling rate needed for complete martensitic microstructures is readily produced through dunk quenching. The workpiece is removed from the induction coil by hand or mechanically and is placed into the quench tank. There are applications, such as the submerged hardening of gear teeth, in which the entire gear is submerged in the quenchant with the induction coil. The gear tooth can be austenitized under the quenchant because the heat formed on the gear tooth produces a vapor pocket during heating of the gear that insulates and permits temperature build-up. At the end of the cycle, turning the power off allows the part to quench in place.

Spray quenches are used both in static heating and single-shot heating, where the workpiece is heated at one time, and in progressive hardening or scanning. The spray quenches can be located either within the coil itself or in an external quench ring. When workpieces are induction scanned, they are moved progressively through the induction coil, heating first for austenization, then quenching as the part passes through the coil. Combinations of dunking the part into a tank, then into a spray quench have been used to produce high cooling rates.

Examples of Basic Quench Methods. Figure 6.10 shows eleven basic arrangements for quenching induction hardened parts:

- The workpiece is removed from the coil for quenching, as shown in Fig. 6.10(a). Some workpieces that are hardened on index tables can be heated in position and then indexed into a spray quench.
- The workpiece is quenched in place from the coil after heating by a spray quench, as shown in Fig. 6.10(b). In this case the coil has two different cooling passages: one on the outside diameter (OD) for cooling the coil with the cooling water on all the time; and the second pas-

sage has the quenchant that is admitted through opening of a quench solenoid at the end of the heat cycle.
- The workpiece is lowered into a spray quench ring at the end of the heat cycle, as shown in Fig. 6.10(c). This is commonly done with lift-and-rotate devices. At the end of the quench cycle, the workpiece may be either lowered or raised for removal.
- The workpiece is lowered into a quench tank for submerged quenching, as shown in Fig. 6.10(d). If oil is used, as in the example, the part must be lowered rapidly and completely into the oil to prevent fire. Polymer may also be used. As previously discussed, polymers will have no fire hazard, but tank temperature control and uniform agitation are much more important. Quench rings can be used in tanks to increase the quenching speed.
- A scanning coil is used, where the quenchant cools the work coil, as shown in Fig. 6.10(e). The quenchant must be flowing during the entire heating cycle.
- A multi-turn scanning coil is shown with a separate quench ring on the exit side of the workpiece, as shown in Fig. 6.10(f). This technique can also be used with single and low-turn coils. One quench ring can be used over a number of different coil diameters. The quench ring can also be spaced at a distance below the coil to permit more equalization of the austenitized temperature before quenching, providing more system versatility.
- The mass of a workpiece is used to self quench the part, as shown in Fig. 6.10(g). A 4340 steel will quench very effectively in some applications using this technique.
- Workpieces are pushed horizontally through an induction coil, dropping into a quench tank to be brought out by a conveyor, as shown in Fig. 6.10(h). Rotary wheel fixtures can use the same technique.
- Vertical scanning in combination with two quenches, as shown in Fig. 6.10(i). When hardening onto the faces of workpieces such as spindles, the cycle initiates with stationary heating of the face, followed by scanning of the body. When the scanner starts to move away from the face, it is necessary to quench the face. A lower quench is used to quench the body during the scanning travel. Sometimes with deep case hardening, a secondary quench is necessary below the primary quench because after the workpiece has passed out of the primary quench, the residual heat in the core can cause the surface to reheat and temper if a secondary quench is not used.
- Vertical scanning where the primary quench is a spray quench from the coil and the secondary quench is in a quench tank, as shown in Fig. 6.10(j). In the figure, the effectiveness of the tank quench is increased through the use of a submerged spray quench.

- A split type coil such as is used in hardening a crankshaft journal with built in spray quench, as shown in Fig. 6.10(k).

Selection of Quenchant

A review of Fig. 6.11 shows the metallurgical basis required to verify the selection of quenchant, that is, to pick a quenchant for the particular part and alloy being cooled that will cool fast enough to permit transfor-

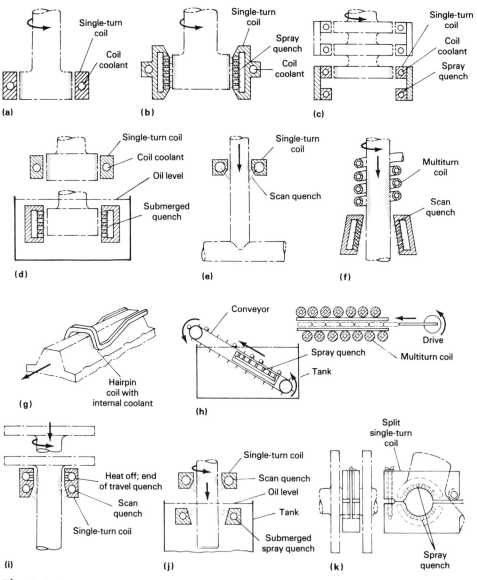

Fig. 6.10 Eleven basic arrangements for quenching induction hardened parts. Source: Ref 1

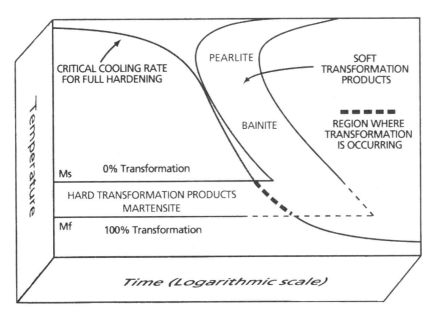

Fig. 6.11 Continuous cooling transformation diagram illustrating the critical cooling rate for complete martensitic transformation. M_s, temperature at which transformation of austenite to martensite starts; M_f, temperature at which transformation of austenite to martensite is completed. Source: Ref 4

mation to martensite without cracking or distortion. Therefore, the general recommendation for selection of a quenchant for induction heat treating is to select the fastest speed quenchant that will produce the desired hardness and case depth without cracking the part. In general with temperature and agitation as a constant, lower carbon steels require the faster cooling rates that are produced by the water or low polymer concentration quenchants. Since the low-carbon steels do not produce quench cracks, the selection of a high-speed quenchant is obvious. Also, fairly fast quenchants are used very effectively in some applications with alloy steel parts where the application involves shallow case depths or small areas to be hardened.

Higher carbon steels, generally over 0.45 carbon may require slower quenchants, and the alloy steels may require high polymer concentrations or oil quenchants. In order to keep alloy steels from cracking, cooling through the M_f to M_s must be uniform. Sometimes testing is needed to determine the use of the in-between concentrations of polymers on specific applications for suitability.

The higher concentrations of polymers have more dragout, adding cost. From a productivity viewpoint, keeping the concentrations as low as practical reduces cost by decreasing the overall cycle time. Finally, the nature of the mechanical handling may help dictate the type of quenchant to be used, while the nature of the part and process may dictate the type of quenchant to be used, while the nature of the part and process may dictate the quenchant concentration and temperature.

REFERENCES

1. *Heat Treating*, Vol 4, *ASM Handbook*, ASM International, 1991
2. *Properties and Selection: Irons and Steels*, Vol 1, *Metals Handbook*, 9th ed., American Society for Metals, 1978
3. *Heat Treating*, Vol 4, *Metals Handbook*, 9th ed., American Society for Metals, 1981
4. E.F. Houghton and Co., "Houghton on Quenching"

CHAPTER 7

Tempering

STEELS ARE TEMPERED to obtain specific values of mechanical properties, to relieve quenching stresses, and to ensure dimensional stability. The highest hardness produced for any given steel without any other surface process is obtained by quenching to a fully martensitic microstructure. A high hardness characteristic of a martensitic microstructure is produced from the straining of the iron lattices by the carbon. Because hardness is directly related to tensile strength, steel composed of 100% martensite is at its strongest possible condition. However, the condition of highest strength is also the condition at which steel is most subject to brittle fracture due to lack of ductility. Except at low carbon contents, fully martensitic steels have insufficient toughness for many applications. Therefore, martensitic steels produced by medium-carbon and high-carbon steels are usually tempered after induction hardening in order to increase their ductility. During tempering, martensitic microstructures supersaturated with carbon are decomposed into a more stable, ductile microstructure. Figure 7.1 shows the increase in ductility (greater reduction in area and more elongation) and lowering of hardness for 4340 steel as it is tempered with increasing temperature. Figure 7.2 shows how increasing the tempering temperature for a 4140 steel lowers the hardness and increases the notch toughness, a measure of ductility.

The variables associated with tempering after induction hardening that affect the microstructure and the mechanical properties include tempering temperature, time at temperature, and the composition of the steel (including carbon content, alloy content, and residual elements). On heating a martensitic microstructure, the carbon atoms diffuse and act over increasing temperature and time in a series of distinct steps that eventually form an iron carbide or an alloy carbide of iron in a ferrite matrix. This gradually decreases hardness, tensile strength, and yield strength, while increasing ductility and toughness. The properties of the tempered steel are determined primarily by the size, shape, composition, and distribution of the carbides that are formed, with a relatively minor contribution from solid-solution hardening of the ferrite. Diffusion of carbon and the alloying

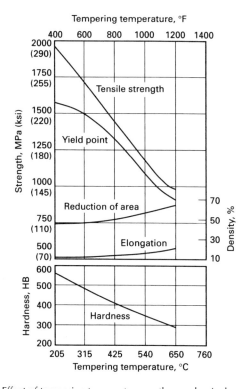

Fig. 7.1 Effect of tempering temperature on the mechanical properties of oil-quenched 4340 steel bar. Single-heat results: ladle composition, 0.41 C-0.67 Mn-0.023 P-0.018 S-0.26 Si-1.77 Ni-0.78 Cr-0.26 Mo; grain size, ASTM 6 to 8; critical points, Ac_1, temperature at which austenite begins to form on heating, 730 °C (1350 °F); Ac_3, temperature at which transformation of ferrite to austenite is completed on heating, 770 °C (1475 °F); Ar_3 475 °C (890 °F); Ar_1 380 °C (720 °F); treatment normalized at 870 °C (1600 °F), reheated to 800 °C (1475 °F), quenched in agitated oil; cross section, 13.46 mm (0.530 in.) diam; round treated, 12.83 mm (505 in.) diam; round tested; as-quenched hardness, 601 HB. Source: Ref 1

elements necessary for the formation of the carbides are temperature and time dependent.

Structural Changes and Stages of Tempering

Based on x-ray diffraction, dilatometric, and microstructural studies, although the temperature ranges may have some overlap, most steels are considered to have three distinct stages of tempering (while some tool steels have a fourth stage).

- *Stage I:* 100 to 250 °C (210 to 480 °F). The formation of transition carbides and lowering of the carbon content of the martensite to 0.25%

- *Stage II:* 200 to 300 °C (390 to 570 °F). The transformation of retained austenite into ferrite and cementite
- *Stage III:* 250 to 350 °C (480 to 660 °F). The replacement of transition carbides and low-temperature martensite by cementite and ferrite
- *Stage IV:* 250 to 700 °C (480 to 1290 °F). The precipitation of finely dispersed alloy carbides and growth in carbide particle size

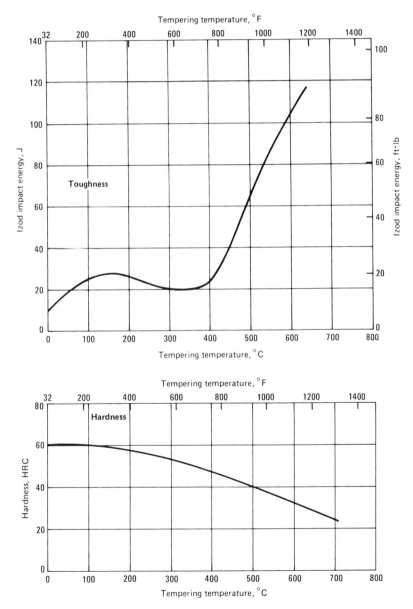

Fig. 7.2 Hardness and notch toughness of 4140 steel tempered 1 h at various temperatures. Source: Ref 2

Dimensional Changes. During tempering, martensite decomposes into a mixture of ferrite and cementite, with a resultant decrease in volume as the tempering temperature increases. Any retained austenite transforms to bainite with an increase in volume in stage II of tempering. When certain alloy steels are tempered, a precipitation of finely distributed alloy carbides occurs, along with an increase in hardness. With the precipitation of these alloy carbides, the M_s temperature of the retained austenite will increase, so that the remaining retained austenite is transformed to martensite during cooling. This tempering, called secondary hardening, does not usually start until over 425 °C (800 °F). After cooling to room temperature, the hardness can be higher than it was as-quenched.

Tempering Temperature

The tempered hardnesses for several furnace-austenitized and quenched steels are presented in Table 7.1. Temperature and time are interdependent variables in the tempering process. Minor temperature changes have more effect than minor time changes. Figure 7.3 shows the decrease in hardness

Table 7.1 Typical hardnesses of various carbon and alloy steels after tempering

Grade	Carbon content, %	\multicolumn{9}{c}{Hardness, HRC, after tempering for 2 h at °C (°F):}	Heat treatment
		205 (400)	260 (500)	315 (600)	370 (700)	425 (800)	480 (900)	540 (1000)	595 (1100)	650 (1200)	
Carbon steels, water hardening											
1030	0.30	50	45	43	39	31	28	25	22	95(a)	Normalized at 900 °C (1650 °F); water quenched
1040	0.40	51	48	46	42	37	30	27	22	94(a)	from 830–845 °C (1525–1550 °F); average dew
1050	0.50	52	50	46	44	40	37	31	29	22	point, 16 °C (60 °F)
1060	0.60	56	55	50	42	38	37	35	33	26	Normalized at 885 °C (1625 °F); water quenched
1080	0.80	57	55	50	43	41	40	39	38	32	from 800–815 °C (1475–1500 °F); average dew
1095	0.95	58	57	52	47	43	42	41	40	33	point, 7 °C (45 °F)
1137	0.40	44	42	40	37	33	30	27	21	91(a)	Normalized at 900 °C (1650 °F); water quenched
1141	0.40	49	46	43	41	38	34	28	23	94(a)	from 830–855 °C (1525–1575 °F); average dew
1144	0.40	55	50	47	45	39	32	29	25	97(a)	point, 13 °C (55 °F)
Alloy steels, water hardening											
1330	0.30	47	44	42	38	35	32	26	22	16	Normalized at 900 °C (1650 °F); water quenched
2330	0.30	47	44	42	38	35	32	26	22	16	from 800–815 °C (1475–1500 °F); average dew
3130	0.30	47	44	42	38	35	32	26	22	16	point, 16 °C (60 °F)
4130	0.30	47	45	43	42	38	34	32	26	22	Normalized at 885 °C (1625 °F); water quenched
5130	0.30	47	45	43	42	38	34	32	26	22	from 800–855 °C (1475–1575 °F); average dew
8630	0.30	47	45	43	42	38	34	32	26	22	point, 16 °C (60 °F)
Alloy steels, oil hardening											
1340	0.40	57	53	50	46	44	41	38	35	31	Normalized at 870 °C (1600 °F); oil quenched
3140	0.40	55	52	49	47	41	37	33	30	26	from 830–845 °C (1525–1550 °F); average dew
4140	0.40	57	53	50	47	45	41	36	33	29	point, 16 °C (60 °F)
4340	0.40	55	52	50	48	45	42	39	34	31	
4640	0.40	52	51	50	47	42	40	37	31	27	
8740	0.40	57	53	50	47	44	41	38	35	22	
4150	0.50	56	55	53	51	47	46	43	39	35	Normalized at 870 °C (1600 °F); oil quenched
5150	0.50	57	55	52	49	45	39	34	31	28	from 830–870 °C (1525–1600 °F); average dew
6150	0.50	58	57	53	50	46	42	40	36	31	point, 13 °C (55 °F)
8650	0.50	55	54	52	49	45	41	37	32	28	Normalized at 870 °C (1600 °F); oil quenched
8750	0.50	56	55	52	51	46	44	39	34	32	from 815–845 °C (1500–1550 °F); average dew
9850	0.50	54	53	51	48	45	41	36	33	30	point, 13 °C (55 °F)

(a) Hardness, HRB. Data were obtained on 25 mm (1 in.) bars adequately quenched to develop full hardness. Source: Ref 2

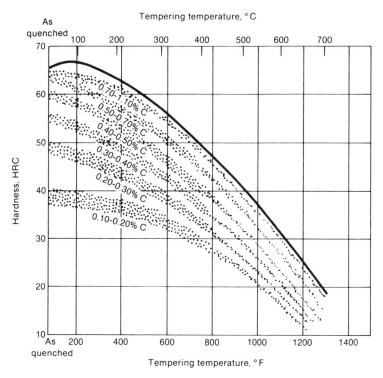

Fig. 7.3 Decrease in hardness with increasing tempering temperature (tempering time, 1 h) for carbon steels of various carbon contents. Source: Ref 3

with increasing tempering temperature for carbon steels of various carbon contents. Induction-hardened parts are normally tempered between 150 to 540 °C (300 and 1100 °F), with most induction-hardened parts probably tempered below 260 °C (500 °F). However, tempering at 150 °C (300 °F) can sometimes drop the hardness below specification. In this case, tempering at 121 °C (250 °F) can produce what is termed *stress relief* and increase the ductility without lowering the hardness.

Tempering Time

The decrease in hardness with time at four tempering temperatures for a 0.82% carbon steel is shown in Fig. 7.4. Time is shown in a logarithmic scale. Most of the hardness decrease occurs in the first hour at heat, while the second hour at heat has a very small decrease. The hardness decrease beyond two hours is negligible. Because of this type of response, tempering temperatures, rather than time, are used as the variable for producing the required hardness. Time is used mainly in three areas. First, continuous furnaces, as described later, need to have the parts in the furnace for minimum times. For example, it may be experimentally verified that 20 min in a

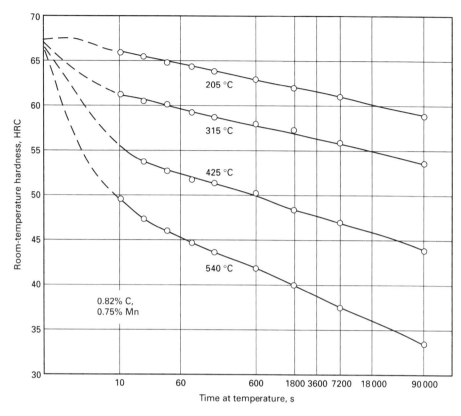

Fig. 7.4 Effect of time at four tempering temperatures on room-temperature hardness of quenched 0.82 C steel. Note nearly straight lines on logarithmic time scale. Source: Ref 1

continuous furnace produces the desired hardness. Second, batch furnaces may have the times at heat increased because of the load density. The mass or weight of the load requires time for the load to uniformly heat to tempering temperature. Typical batch loads run at heat from 1 to 2 h. Finally, induction tempering requires sufficient time for heat transfer in the workpiece, so that the entire hardened area is heated to tempering temperature. Typical tempering times for induction tempering may run from 10 to 20 s.

Tempering Processes

There are three common types of tempering processes that are used after induction hardening a workpiece: furnace tempering, induction tempering, and residual heat tempering.

Furnace tempering is done with either batch or continuous ovens.

Batch tempering is most commonly done in an oven. Steel workpieces are loaded into baskets or retorts, so that the entire load is tempered in one heat at the temperature and time required for lowering the surface hardness into the specified range. The most uniform heating in batch ovens is accomplished with ovens that have fans to provide high air recirculation through the load. The heating characteristics of the furnace being used should be considered when selecting the time to use for tempering. Load density and convection in the furnace change the time required to bring the entire load up to temperature. Depending on the load density and intensity of air recirculation and type of oven, the normal rule for tempering time is from one to two hours at heat. Once the heating time cycles are established, tempering temperature is normally used to determine the required change in hardness. The tempering curves for several commonly induction hardened 0.45% carbon steels are shown in Fig. 7.5. These tempering curves

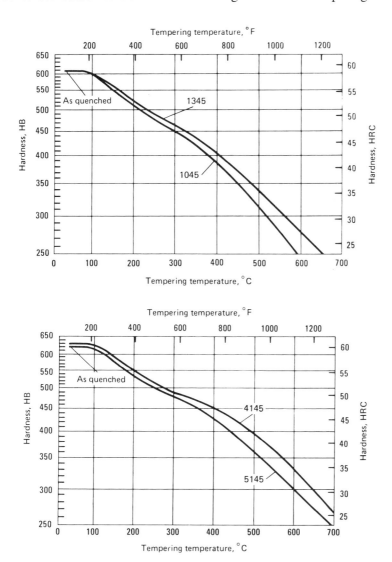

Fig. 7.5 Tempering characteristics of four 0.45 carbon and alloy steels tempered 1 h. Source: Ref 2

are used as the model for selection of tempering temperature by batch furnace tempering. As previously mentioned, there is usually no drop in hardness in induction-hardened workpieces when tempered at 121 °C (250 °F). When workpieces quench harder or softer than the curves, interpolate using the drop in hardness with change in temperature to calculate a revised tempering temperature. For instance, if a 4145 steel workpiece quenches to 58 HRC and needs to be tempered to 50 to 55 HRC, a 4.5 point drop in hardness requires a tempering temperature of about 205 °C (400 °F). A workpiece can be run through on a pilot test to verify the temperature that is correct for the hardness drop. Then, if the hardness is off by only a point or two, a minor adjustment in temperature is needed. Carbon steels have a nearly linear decrease in hardness over 149 °C (300 °F), with a drop of one point HRC for roughly every 13 °C (25 °F) increase in temperature. This rule is useful when there are variations in the as-quenched hardness, and minor adjustments need to be made in the tempering temperature. Caution must be used when tempering alloy steels. Many of the textbook curves may not show the decrease in hardness that may occur below 204 °C (400 °F), and the decrease in hardness may not be linear. Additional tempering curves for steels commonly induction hardened are shown in Appendix "Tempering Curves" in this book.

Continuous furnaces use higher tempering temperatures to reduce the as-quenched hardness than batch furnaces because the workpieces are not held at temperature as long. Because of this, the time and temperature relationships for the needed drop in hardness have to be more closely reviewed before processing. Again, the tempering of pilot workpieces is recommended.

Finally, if workpieces are run through at a lower temperature that does not produce the necessary drop in hardness, the workpieces can be retempered at a higher temperature without any detrimental effect on the properties. For example assume that a specification calls for HRC 51 to 55. If a load of 0.45% carbon steel is tempered at 204 °C (400 °F) to produce a final hardness of HRC 55, the load can be rerun at 232 °C (450 °F) to produce an expected hardness of HRC 53. The drop in hardness reacts to the temperature of the tempering operation in the second heating cycle, almost as if the first lower tempering operation had not been done.

Induction Tempering. The major differences between induction and furnace tempering are the times and temperatures involved. When workpieces are tempered by induction, they are heated to a given surface temperature with no holding time at temperature. Figure 7.6 shows hardness as a function of tempering temperature and time, for furnace- and induction-treated 1050 steel. The shorter cycle of an induction-tempered workpiece requires higher tempering temperatures. In Fig. 7.6 the austenitized and quenched hardness of the steel was 62 HRC. A 1 h furnace tempering

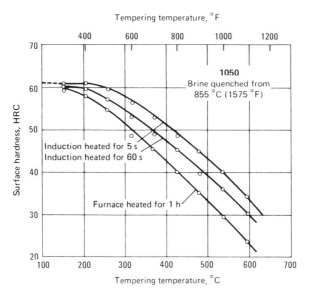

Fig. 7.6 Variation of hardness with tempering temperature for furnace and induction heating. Source: Ref 4

cycle at 425 °C (800 °F) produced the same 40 HRC hardness as the same steel heated for 5 seconds to 540 °C (1000 °F). The faster tempering cycle requires a higher temperature in order for diffusion of the carbon atoms to form iron carbide. Because other mechanical properties, such as yield and tensile strength, elongation, reduction in area, and fracture toughness, often correlate with hardness, these two different treatments produce similar products. However, the mechanical properties may be slightly lower in induction-tempered workpieces than in furnace tempered workpieces, because of the lack of time at temperature and the slight differences in microstructure and residual stress.

Tempering always involves a reasonable compromise between maintaining the required hardness and obtaining a low-stress, tough, and ductile microstructure. Because stress relief is an important goal of tempering, it is imperative to know how internal stresses are relieved during induction tempering. Prior macrostresses, thermal stresses, and transformation stresses were discussed in the chapter "Heat Treating Basics" in this book. The overall macrostresses may not be the same, and the microstructure may not be totally identical when comparing induction tempering to furnace tempering. Figure 7.7 shows the temperature-versus-time plot for the induction tempering of a bar. The temperature distribution over heating and cooling is not as uniform as that produced during furnace tempering. Therefore, it is important that the mechanical properties be tested when setting up an induction tempering line.

The controls needed for induction tempering must be very stringent in terms of both time and temperature. The consistency of producing the

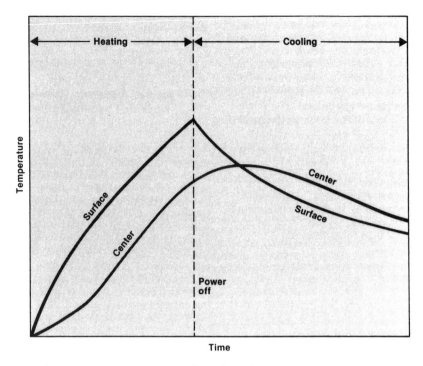

Fig. 7.7 Temperature vs. time plot at the surface and center of an induction-heated, solid bar. Source: Ref 5

same tempered hardness needs to be examined from all process variables. This includes variables such as different heats of steel, the effects on non-uniform geometry of the workpieces, and the ability of the induction heating system to provide precisely the same amount of power into the workpiece throughout the day.

The general recommendation, where possible, is that a workpiece that is outside diameter (OD) induction hardened should be inside diameter (ID) induction tempered. Conversely, ID induction-hardened workpieces should be OD induction tempered. The reason for this is the shallow reference depth that is produced in steel under the Curie Temperature, as previously described in the Chapter "Theory of Heating by Induction" in this book. The initial high heat build-up is in the reference depth, which is usually less than the case or cross section, being tempered. At tempering temperatures, the rate of heat conduction is relatively slow, and it is harder to provide uniform tempering temperatures with the short heating time throughout the martensitic area. By heating from the opposite side, where higher temperatures have no effect on the properties of the workpiece, the heat flow provides a more gradual build-up of the temperature in the martensitic area and provides more uniform tempering temperature. Because of the slower rate of heat conduction, it is not unusual for the induction tempering heating time to exceed twice the austenitizing time for the same workpiece. An induction-hardening process may have an indexing fixture with one heating position and frequency for austenitizing and quenching, and two other positions of a lower frequency for tempering. The heating time for an individual tempering position is equal to the austenitizing time. The total time for tempering is the total of the time spent in both tempering positions and the index in-between. Through a process such as this, the effective production rate as produced by the time of the austenitizing cycle is not reduced by the requirement for longer tempering time.

Additional care must be taken with regard to the selection of frequency, coil design, and technique of heating. Table 7.2 provides information regarding selection of power source and frequency for various applications of induction tempering. The geometry of the workpiece obviously must be considered to recognize the effect of thermal mass on the induced energy. For instance, in the case of induction tempering of gears, it is very important that the tooth and the root of the tooth have identical heating time-temperatures.

Equivalent time/temperature cycles for tempering are determined in several ways. For example, tempering experiments can be conducted at a variety of temperatures for a range of times. The relationship for induction is determined and then plotted to establish the time-temperature relationship for the workpieces being heated. Table 7.3 provides operating and production data for progressive induction tempering.

A number of researchers have found that the hardness of tempered martensite can be correlated to a certain function of the tempering time and tempering temperature. A simple technique uses a mathematical function known as the tempering parameter (*T.P.*) which was developed to provide an approximation of the tempering temperatures. The function is $T.P. = T(14.44 + \log t)$ where T is the tempering temperature in degrees Rankine, 14.44 is a constant for carbon steels, and t is the tempering time in temperature. Using this formula, the results show that furnace tempering for 1 h at 425 °C (800 °F) provides the same hardness as tempering by

Table 7.2 Selection of power source and frequency for various applications of induction tempering

Section size		Maximum tempering temperature		Rating(a) for:					
				Power lines,	Frequency converter,	Solid-state systems or motor-generators			Vacuum tube over
cm	in.	°C	°F	50 or 60 Hz	180 Hz	1000 Hz	3000 Hz	10,000 Hz	200 kHz
0.32 to 0.64	1/8 to 1/4	705	1300	...	...	...	...	...	Good
0.64 to 1.27	1/4 to 1/2	705	1300	...	...	...	...	Good	Good
1.27 to 2.54	1/2 to 1	425	800	...	Fair	Good	Good	Good	Fair
		705	1300	...	Poor	Fair	Good	Good	Fair
2.54 to 5.08	1 to 2	425	800	Fair	Fair	Good	Good	Fair	Poor
		705	1300	...	Fair	Good	Good	Fair	Poor
5.08 to 15.24	2 to 6	425	800	Good	Good	Good	Fair	...	...
		705	1300	Good	Good	Good	Fair	...	...
Over 15.24	Over 6	705	1300	Good	Good	Good	Fair	...	...

(a) Efficiency, capital cost, and uniformity of heating are considered in these ratings. "Good" indicates optimum frequency. "Fair" indicates frequency higher than optimum that increases capital cost and reduces uniformity of heating, thus requiring lower heat inputs. "Poor" indicates frequency substantially higher than optimum that substantially increases capital cost and reduces uniformity of heating, thus requiring substantially lower heat inputs. Source: Ref 3

Table 7.3 Operating and production data for progressive induction tempering

Section size		Material	Frequency, Hz	Power(a), kW	Total heating time, s	Scan time		Work temperature				Production rate		Inductor input(b)	
								Entering coil		Leaving coil					
cm	in.					s/cm	s/in.	°C	°F	°C	°F	kg/h	lb/h	kW/cm^2	kW/in.^{2}
Rounds															
1.27	½	4130	9600	11	17	0.39	1	50	120	565	1050	92	202	0.064	0.41
1.91	¾	1035 mod	9600	12.7	30.6	0.71	1.8	50	120	510	950	113	250	0.050	0.32
2.54	1	1041	9600	18.7	44.2	1.02	2.6	50	120	565	1050	141	311	0.054	0.35
2.86	1⅛	1041	9600	20.6	51	1.18	3.0	50	120	565	1050	153	338	0.053	0.34
4.92	1¹⁵⁄₁₆	14B35H	180	24	196	2.76	7.0	50	120	565	1050	195	429	0.031	0.20
Flats															
1.59	⅝	1038	60	88	123	0.59	1.5	40	100	290	550	1449	3194	0.014	0.089
1.91	¾	1038	60	100	164	0.79	2.0	40	100	315	600	1576	3474	0.013	0.081
2.22	⅞	1043	60	98	312	1.50	3.8	40	100	290	550	1609	3548	0.008	0.050
2.54	1	1043	60	85	254	1.22	3.1	40	100	290	550	1365	3009	0.011	0.068
2.86	1⅛	1043	60	90	328	1.57	4.0	40	100	290	550	1483	3269	0.009	0.060
Irregular shapes															
1.75 to 3.33	¹¹⁄₁₆ to 1¹⁵⁄₁₆	1037 mod	9600	192	64.8	0.94	2.4	65	150	550	1020	2211	4875	0.043	0.28
1.75 to 2.86	¹¹⁄₁₆ to 1⅛	1037 mod	9600	154	46	0.67	1.7	65	150	425	800	2276	5019	0.040	0.26

(a) Power transmitted by the inductor at the operating frequency indicated. For converted frequencies, this power is approximately 25% less than the power input to the machine, because of losses within the machine. (b) At the operating frequency of the inductor. Source: Ref 3

induction to 540 °C (1000 °F) in 5 s. It should be noted that these calculations assume the workpiece has uniform geometry with good heat transfer.

Residual heat tempering uses heat in the core of the workpiece to raise the surface heat back to tempering temperature after the surface was quenched below the M_f temperature. The process must be designed to produce the same temperature on quenching, so that the workpiece can use the residual heat to bring the surface temperature back up into the predetermined tempering range. Upon cooling, the microstructure will be tempered martensite. This process is application specific to certain products such as larger diameter bars and pins that have sufficient heat in the core to bring the surface temperature into tempering range.

REFERENCES

1. *Heat Treating*, Vol 4, *ASM Handbook*, ASM International, 1991
2. *Properties and Selection: Irons and Steels*, Vol 1, *Metals Handbook*, 9th ed., American Society for Metals, 1978
3. S.L. Semiatin and D.E. Stutz, *Induction Heat Treating of Steel*, American Society for Metals, 1986
4. *Heat Treating*, Vol 4, *Metals Handbook*, 9th ed., American Society for Metals, 1981
5. Tech Commentary, "Induction Tempering," Vol 2, No. 4, Center for Metals Fabrication, 1985

CHAPTER **8**

Cleaning and Rust Protection

WORKPIECES MAY NEED cleaning both before and after induction hardening. Machining fluid residues not only smoke during austenitizing, but they can also severely contaminate quenchants. This contamination will cause the quenchant's cooling rate to change, which of course is an undesirable condition. Furthermore, experience has shown that the workpieces need to be cleaned before furnace tempering. Finally, after tempering, unless the workpieces are processed in line in a manufacturing cell, some sort of rust protection is advisable. With in-line processes, the necessity for cleaning and rust protection depends on the other in-line processes that will be done.

Cleaning

Different types of quenchants have different types of cleaning requirements before tempering. Oil quenchants need to be removed before tempering. Quench oils burn in the tempering furnaces, and if they are not removed, the parts in a tempering retort can actually glue together. The oil can coat the parts, similar to a thin lacquer, making cleaning very difficult after removal from the tempering furnace. While many types of solvent cleaners can be used, environmentally it is easy to use an industrial washer to remove the oil.

Water quenches will leave the parts with a tendency to rust unless chemical rust inhibitors are used in the quench water. Sometimes enough residual heat can be left in the parts that any remaining water evaporates. Under continuous conditions, air blasts may be able to be used to remove the water. Rotating spindles with workpieces can have a rotation speed high enough to spin off the water.

Polymer quenches present a different situation. While the manufacturers state in their literature that parts can be tempered without washing, experience shows that the polymers must be rinsed or washed before tempering. Several things can happen when parts are tempered without polymer removal. First, evaporation and burning of the polymer in the tempering process create obnoxious odors. Second, the polymers form adherent coatings that are difficult to remove after tempering. Third, polymers seem to make parts more susceptible to rusting at a later date—even if rust preventives are applied after tempering.

Rust Protection

There are a number of rust preventives that can be applied for temporary protection. Keep in mind that when removed from the tempering furnace, the parts have clean surfaces. Rust can form very quickly, particularly in humid weather. Even if not seen, rust that is not visible can form in the pores. Over the period of several months, the rust will nucleate and grow. Therefore, it is recommended that rust preventives be applied to the parts as soon as possible after they have cooled.

CHAPTER 9

Decarburization and Defects

DECARBURIZATION may or may not be a defect on a workpiece. The potential for decarburization needs to be understood so that workpieces are not rejected for low hardness during inspection. The intent of this Chapter is to explain decarburization and the importance of recognizing decarburization when induction hardening. Also, the nature and origin of various defects are explained so that proper analysis can be used to identify defects, both before and after induction hardening, that may make workpieces nonconforming.

Decarburization

Decarburization is a surface condition that is common to all hot-rolled steel products to some extent. It is produced during heating and rolling operations when oxygen in the atmosphere at temperature reacts with and removes carbon from the hot surface. Decarburization can also be produced when parts are furnace austenitized in air or incorrect protective atmospheres. The depth of decarburization depends on temperature, time at temperature, nature of furnace atmosphere, reduction of area between the bloom and the finished size, and the type of steel. Figure 9.1 shows several examples of decarburization. Figure 9.1(b) shows surface decarburization and the increase in surface hardness, with hardness readings taken from the surface inwards. Figure 9.1(c) shows the completely ferritic surface layer in a commercial 1.0% carbon strip in the austenitized and quenched condition. From an induction heat treating viewpoint, any time plate or strip steel is being used as the material for a workpiece, there is potential for surface decarburization. Figure 9.2 shows the decarburization in a normalized, 0.8% carbon steel. Picral etchants define pearlite and show the ferrite (the decarburized area) as white. The nital etchants that are used for

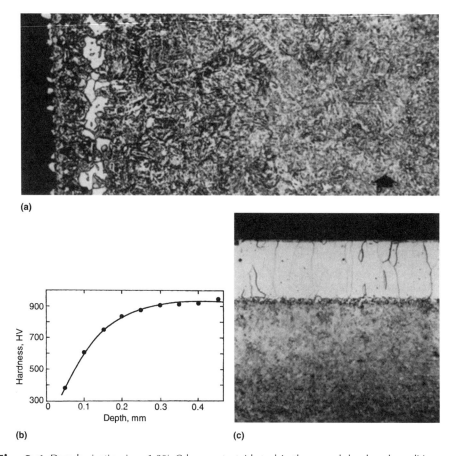

Fig. 9.1 Decarburization in a 1.3% C hypereutectoid steel in the quench-hardened condition. 1.20 C-0.17 Si-0.40 Mn. Austenitized at 850 °C (1562 °F), water quenched, tempered at 175 °C (347 °F). (a) 1% nital. 250×. (b) Variation of hardness with depth in the quenched-and-tempered hypereutectoid steel. (c) Completely ferritic surface layer in a commercial 1% C strip in the quench-hardened condition. 3% nital. 250×. Source: Ref 1

defining induction-hardened cases will not define decarburization very well unless the workpieces are tempered at a sufficiently high temperature to produce dark, tempered martensite when etched. Decarburization lowers surface hardness, producing a soft, low-strength area and permits plastic deformation at the root of a notch without crack initiation. Thus it will not lower, but may slightly increase, notch toughness. Decarburization also lowers the fatigue resistance of steel. Figure 9.3 illustrates the effect of decarburization on fatigue behavior of a steel. The decarburized parts have average lower fatigue lives. For this reason, critical parts, or at least critical areas of parts with decarburized surfaces, are usually machined to remove the decarburized material.

Depth of Decarburization Produced by Induction. For a typical 5 s heat cycle to 950 °C (1760 °F), the decarb depth is calculated to be

Decarburization and Defects / 153

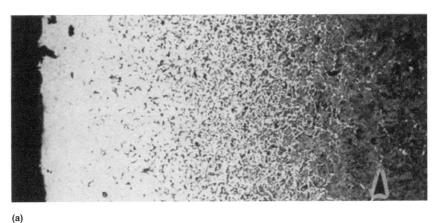

(a)

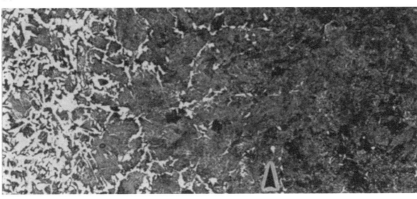

(b)

Fig. 9.2 (a) Decarburization in a 0.8% C eutectoid steel, 0.78 C-0.30 Mn. (b) Transverse section of a hot-rolled bar; normalized. Arrows indicate total depth of decarburization. (a) Picral. 50×. (b) Picral. 100×. Source: Ref 1

0.00197 cm (0.00078 in.). This depth is so shallow that decarburization with induction heat treating has never been a problem. Castings, forgings, cold-drawn stock, and hot-rolled products may all have decarburization unless the decarburization was removed, or unless there was a carbon restoration heat treating process. Another source of decarburization occurs when parts are heat treated in nonprotective atmospheres or in atmospheres in which the carbon potential is such that decarburization is not prevented during austenitizing.

Castings. Decarburization is a normal product found on the surface of castings. Even investment castings, with the close finished tolerances, may have decarburization.

Forgings. Parts before forging are normally heated to forging temperatures without the benefit of protective atmospheres. Therefore, unless a carbon restoration process has been done, the forgings are likely to have decarburization. Table 9.1 shows the amount of stock removal recommended

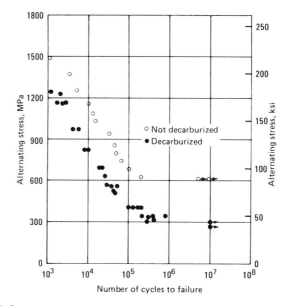

Fig. 9.3 Effect of decarburization on fatigue behavior of a steel. Source: Ref 2

for machining in highly stressed areas. Low surface hardness may result unless all the decarburization is removed.

Stock Removal for Cold Finished Steel. Standard quality cold-finished bars are produced from hot-rolled steel. Therefore, the original decarburization that was present on the hot-rolled stock is still present on the steel, although it has been reduced in thickness as the bar diameter was reduced by cold drawing. Guidelines provide recommendations for stock removal of surface decarburization and any other surface defects that might be present. The recommended stock removal for all nonresulfurized grades is 0.025 mm per 1.59 mm (0.001 in. per 1/16 in.) of cross section, or 0.254 mm (0.010 in.), whichever is greater. For example, for a 25-mm (1-in.) bar, recommended stock removal is 0.254 mm (0.016 in.) per side. For the resulfurized grades, recommended stock removal is 0.038 mm per 1.59 mm (0.0015 in. per 1/16 in.) or 0.38 mm (0.015 in.), whichever is greater.

Stock Removal for Hot-Rolled Stock. Stock removal of 3.18 mm (1/8 in.) is recommended for diameters of 38 through 76 mm (1½ through 3 in.) and 6.35 mm (¼ in.) of larger diameter bars.

Table 9.1 Typical decarburization limits for steel forgings

Range of section size		Typical depth of decarburization	
mm	in.	mm	in.
<25	<1	0.8	0.031
25 to 100	1 to 4	1.2	0.047
100 to 200	4 to 8	1.6	0.062
>200	>8	3.2	0.125

Source: Ref 2

Significance on Induction Hardening. Prints, as provided for parts that are to be induction hardened, do not generally indicate whether sufficient stock has been machined to remove any potential surface decarburization. The induction-hardening process by itself is not affected by decarburization. However, the lower surface hardness produced after quenching can cause the quenched workpieces to be in nonconformance. The Chapter "Nonconforming Product and Process Problems" in this book presents methods of checking for decarburization. Proper product design must make allowances for removal of decarburization, or carbon restoration of decarburization. It should be noted that in some instances in which there are finishing operations after induction hardening, stock removal can be completed after heat treating. In these cases surface hardness specifications must allow for the soft hardnesses produced by decarburization.

Defects and Flaws

The objective of induction heat treating is to produce defect-free workpieces that are hardened to specifications. However, there are defects that can be present in the workpieces before induction hardening. These defects may cause problems that no amount of proper heat treating technique can overcome. There are also defects that can be produced by the induction heat treating process itself.

By definition, a defect is a discontinuity the size, shape, orientation, or location of which makes it detrimental to the useful service of the workpiece in which it occurs. Flaws are a detectable imperfection in a workpiece. A flaw may not make a workpiece nonconforming. Fractures initiate at localized nonhomogenities or defects in the material such as inclusions, microcracks, and voids. Defects and flaws are detrimental if they lead to stress concentrations that cause a fracture anytime in the manufacturing cycle or life cycle of a part. The different manufacturing processes for workpieces that are induction hardened have different types of defects.

Defects or flaws most likely to be found in workpieces before induction hardening are defects such as casting defects, cold heading defects, and defects in bar stock. Figure 9.4 shows ten different types of flaws that may be found in rolled bars.

- *Porosity:* the result of trapped gaseous bubbles in the solidifying metal, causing porous structure in the interior of the ingot. On rolling, these structures are elongated and interspersed throughout the cross section of the bar product as illustrated. The major casting imperfections that affect induction hardened workpieces are blowholes or gas holes which may appear as subsurface voids. These voids can cause nonuniform heating and hot spots.

- *Inclusions:* (Fig. 9.4a) may occur from spatter (entrapped splashes) during the pouring of the steel into the mold. They are elongated during rolling and are usually subsurface in the bar, showing a lamellar structure of inclusions which is seen in the longitudinal or rolling direction. The inclusions can be oxide in nature, or they can be from sulfur that is added to form manganese sulfide for machinability. Figure 9.5 shows a micrograph of manganese sulfide inclusions.
- *Chips:* (laminations, Fig. 9.4b) occur from spatter (entrapped splashes) during the pouring of steel into a mold
- *Slivers:* (Fig. 9.4c) most often are caused by a rough mold surface, overheating before rolling, or abrasion during rolling. Usually slivers are found with seams.
- *Embedded scab:* (Fig. 9.4d) caused by metal that is removed from the wall of the mold, which becomes embedded in the surface of the rolled bar
- *Pits and blisters:* during subsequent rolling, gaseous pockets in the ingot often become pits or blisters (Fig. 9.4e) on the surface or slightly below the surface of bar products.
- *Embedded scale:* (Fig. 9.4f) may result from rolling or drawing of bars that have become excessively scaled during prior heating operations
- *Cracks and seams:* often confused with each other. Figure 9.4(g) shows an example of a crack. Cracks with little or no oxide present on their edges may occur when the metal cools in the mold, setting up highly stressed areas. Seams that develop from these cracks during rolling are heavily oxidized. Cracks also result from highly stressed planes in cold-drawn bars, or from improper quenching during heat treatment. Cracks created from these latter two causes show no evidence of oxidized surfaces (Fig. 9.4g). The defects that show up in fasteners are usually cracks caused by overstressing the metal during heading or other forming operations. Typical locations of forging cracks from cold forging are shown in Fig. 9.6.
- *Seams:* result from elongated, trapped gas pockets or from cracks. The surfaces generally are heavily oxidized and decarburized. Depth varies widely, and surface areas sometimes may be welded together in spots. Seams may be continuous or intermittent, as indicated in Fig. 9.4(h). A micrograph of a typical seam is shown in Fig. 9.7.
- *Laps:* most often caused by excessive material in a given hot roll pass being squeezed out into the area of the roll collar. When turned for the following pass, the material is rolled back into the bar and appears as a lap on the surface (Fig. 9.4j), somewhat similar to seams that result from rolling fins or protrusions into the surface of the bar. The net effect is about the same as far as imperfections are concerned.
- *Chevrons:* (Fig. 9.4k) internal flaws named for their shape. They often result from excessively severe cold drawing and are even more likely

to occur during extrusion operations. The severe stresses that build up internally cause transverse subsurface cracks. Most of the bar stock sold will have the seams removed either during manufacturing or during machining. The stock removal, as mentioned earlier in this chapter for the removal of decarburization, generally removes all seams and laps. However, occasionally seams are found in bar stock that is used to make workpieces for induction hardening.

Surface Defect and Seam Detection. Sometimes surface defects can be seen visually without magnification. When a defect is found, workpieces should be inspected before induction processing. Seams appear as fine lines or cracks that extend longitudinally. There may be more than one seam around the diameter, and the seams may be discontinuous, appearing and disappearing down the length of the workpiece. Most of the time the seams are shallow. Seams and cracks smaller than approximately 0.013 mm (0.005 in.) can be detected through the use of die penetrants as used with fluorescent light. Larger defects can be detected by magnafluxing.

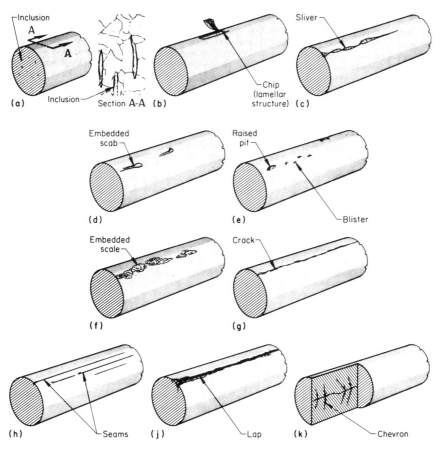

Fig. 9.4 Ten different types of flaws that may be found in rolled bars. Source: Ref 3

158 / Practical Induction Heat Treating

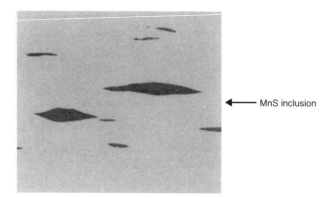

Fig. 9.5 0.39 carbon steel, unetched, 250×. Source: Ref 1

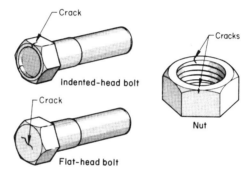

Fig. 9.6 Typical locations of forging cracks in bolt heads and in nuts. Source: Ref 3

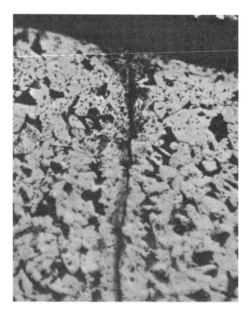

Fig. 9.7 Micrograph of a seam in cross section of a ³⁄₄ in. diam medium-carbon steel bar, showing oxide and decarburization in the seam. 350×. Source: Ref 3

Another method for seam detection is visual observation of the heated area during induction austenitizing. The current concentration produced by the eddy currents flowing perpendicular to a seam will cause the seam itself to heat hotter than the balance of the workpiece. The seam appears as a hot, white line inside the induction coil. Operators can be trained to watch for this.

Seams, when detected after heat treating, have the appearance of a heat treating crack. At this point it can be difficult to visually determine whether the crack is a true heat treating crack or a seam. Seams can initiate quench cracks that extend deeper than the initial depth of the seam. If a crack extends into a ductile area that is not heat treated, a seam is usually indicated. However, with some of the high tensile strength steels, cold drawn steel seams can initiate a heat-treating crack that extends the full length of the workpiece. As previously stated, the best time for positive detection and identification of seams is before induction hardening.

Stress and Quench Cracks

Stress cracks can be produced in cold-drawn bars that crack during heating up to austenitizing temperature. High nonuniform residual stresses, as discussed in the Chapter "Heat Treating Basics" in this book, can cause the workpiece to crack when the hot surface cracks during heating due to the expansion caused by high stresses toward the core with resulting lower yield strength.

Quench cracks occur after a workpiece is heated and quenched. Quench cracks initiate most commonly in the hardened case when the brittle martensitic case cracks due to stresses produced in the later transformation of austenite to martensite under the case. Quench cracks can also initiate at stress risers, such as keyways, holes, or fillets. Quench cracking is mostly intergranular (along the grain boundaries). There have been instances of cracking at the boundary of the induction-hardened case and the unhardened core. The main reasons for cracking in heat treatment are workpiece design, grades of steel used, workpiece defects, heat treating practice, and tempering practice.

Workpiece Design. Features such as sharp corners, fillets, radii, the number, location and size of holes, toolmarks, and keyways are termed *stress risers*. Abrupt changes in section thickness within a workpiece can promote nonuniform cooling during quenching and can initiate quench cracks in the section that hardens first. Wherever possible workpieces should be designed to reduce the stress risers that initiate quench cracks. Radii and corners should be rounded. Holes should have chamfer relief, and keyways should not be hardened unless necessary. When hardening of keyways is necessary, careful selection of the induction power supply frequency is required. The quench selected needs to cool the quenched area uniformly as slowly as possible. Sometimes holes and keyways in bores can be machined after the workpieces are induction hardened.

Workpiece Defects. Surface defects or weaknesses in the material may also cause cracking. This includes defects such as subsurface oxides or inclusions, internal ruptures in forgings, and casting defects.

Heat Treating Practice. Excessive austenitizing temperatures increase the propensity toward formation of quench cracks. Similarly, steels with coarser grain size are more prone to cracks than fine-grain steels because the fine-grain steels possess more grain-boundary area to stop the movement of cracks and to redistribute residual stresses. Figure 9.8 shows examples of several types of quench cracks.

Medium-carbon steels are often quenched with water or low polymer-water quenchants to produce high hardness, and they are susceptible to cracks if they are not handled properly. At high hardness levels martensite is very brittle and can easily crack on quenching, particularly when the workpiece is not uniform. Another problem occurs with design tendencies that try to keep the carbon content low to increase machinability. This can result in needing faster quenchants to produce the specified hardness, leading to quench cracking. One steel that is very prone to producing a quench crack when a quenched hardness of over 57 HRC is needed is 4140.

When quench cracking occurs, the workpiece and crack should be examined very carefully. If workpieces are available that have not been heat treated yet, they should be examined for flaws that might be present. If possible, magnaflux the workpieces and then harden them. If quench cracks are found after magnafluxing, then the likelihood is that the cracking is occurring during quenching. The cracks may need to be examined microscopically to look for subsurface defects and flaws. In order to eliminate the cracking, the very first thing to do is to improve the quenching. A slower-speed quenchant should be selected. The quenching speed can be slowed only to the point that there is a loss of either case depth or surface hardness. When deep case quenching, other factors need to be examined. Sometimes the workpieces need to be held in the quenchant longer so that the heat is extracted from the core faster. Workpieces that are scanned can need secondary quenching applied (another quench ring applying quenchant below the first area. Sometimes substantial testing of various quenchants and quenching rates is necessary. There have been instances in which a faster quench stopped the

Decarburization and Defects / 161

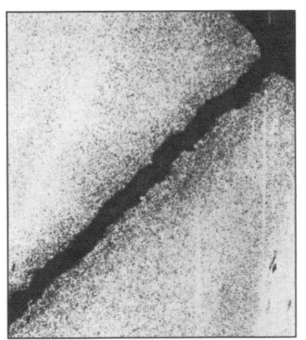
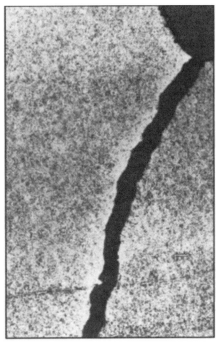
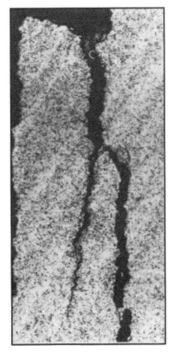
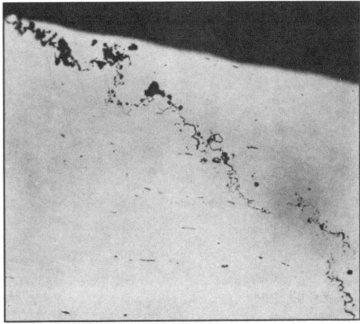

(a) Society of Automotive Engineers (SAE) 4140 steel as-quenched and tempered; microstructure is tempered martensite with quench crack at area of dimensional change. 2% Nital. 100×

(b) SAE 4142H steel as-quenched and tempered; microstructure is tempered martensite with quench crack at the radius. 3% Nital. 100×

(c) SAE 4150 steel as-quenched and tempered; cracking initiates from silicate and sulfide inclusions. 2% Nital. 100×

(d) SAE 4140 steel as-quenched and tempered; microstructure is tempered martensite with cracking at inclusions. Unetched. 100×

Fig. 9.8 Micrographs of cracks. Source: Ref 4

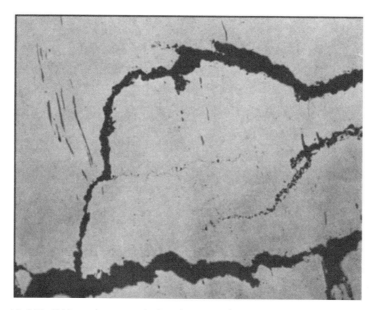

(e) SAE 4140 steel as-quenched and tempered; microstructure is tempered martensite with quench cracking promoted by nonmetallic inclusions. Unetched. 100×

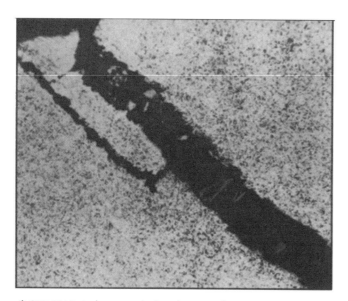

(f) SAE 1144 steel as-quenched and tempered; microstructure is tempered martensite where cracking is aided by inclusion defects. 2% Nital. 200×

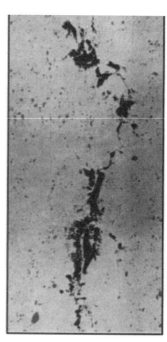

(g) SAE 1144 steel as-quenched and tempered; microstructure is tempered martensite where cracking is aided by inclusion defects. Unetched. 100×

Fig. 9.8 (continued) Micrographs of cracks. Source: Ref 4

Decarburization and Defects / 163

(h) SAE 8630 steel as-quenched; microstructure is martensite where cracking initiated from rolling seam.

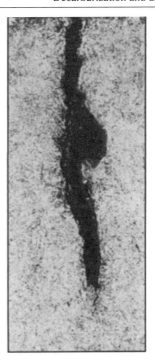

(i) SAE type 403 stainless steel as-quenched and tempered; microstructure is predominantly tempered martensite with cracking promoted by the seam. Vilellas. 100×

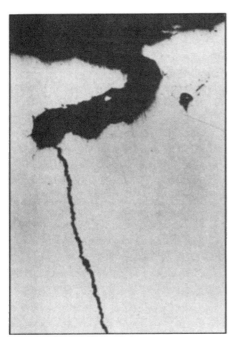

(j) SAE 4118 carburized steel as-quenched and tempered; microstructure is tempered martensite with quench crack propagating from machine burr. Unetched. 200×

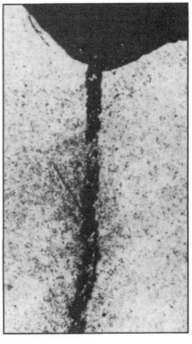

(k) SAE 4140 steel as-quenched and tempered; microstructure is tempered martensite with quench crack initiating from machine groove. 2% Nital. 100×

Fig. 9.8 (continued) Micrographs of cracks. Source: Ref 4

cracking, apparently by producing a deeper martensitic case that had sufficient strength to withstand the stresses produced by the deeper core transformation to martensite.

Good metallurgical practice has always been to temper quenched workpieces as soon as possible. Immediate tempering reduces the martensite brittleness and stops quench cracking. Because immediate tempering is difficult for workpieces that are batch tempered, the best practice is to temper as soon as possible, at least in the same shift. The longer steel workpieces are kept at room temperature after transformation to martensite, assuming that the M_f temperature has not been reached, the more likely the occurrence of quench cracking from the volumetric expansion caused by isothermal transformation of retained austenite into martensite. This is more of a process problem with some of the alloy and tool steels that can have considerable retained austenite.

Finally, workpieces that are induction tempered should have the cased area cooled below the M_f before being reheated for tempering. If workpieces are reheated before passing the M_f temperature, the final microstructure should be reviewed for complete transformation to tempered martensite.

REFERENCES

1. L.E. Samuels, *Light Microscopy of Carbon Steel*, ASM International, 1999
2. *Properties and Selection: Irons and Steels*, Vol 1, *Metals Handbook*, 9th ed., American Society for Metals, 1978
3. *Nondestructive Inspection and Quality Control*, Vol 11, *Metals Handbook*, 8th ed., American Society for Metals, 1976
4. R.R. Blackwood and Larry M. Jarvis, *Ind. Heat*, reprint from Tenaxol Corp.

CHAPTER **10**

Applications of Induction Heat Treatment*

SINCE ITS INTRODUCTION in the 1930s, induction heat treatment has been applied to a large variety of mass-produced commercial products. The initial applications involved hardening of the surfaces of round steel parts such as shafts. Subsequent surface-hardening techniques were developed for other parts whose shapes are not so simple. Most recently, induction hardening and tempering techniques have been developed for purposes of heat treating to large case depths and heat treating entire cross sections. Types of parts to which induction is commonly applied include the following:

Typical parts induction surface hardened include:

- *Transportation field:* crankshafts, camshafts, axle shafts, transmission shafts, splined shafts, universal joints, gears, valve seats, wheel spindles, and ball studs
- *Machine-tool field:* lathe beds, machine beds, transmission gears, and shafts
- *Metalworking and hand-tool fields:* rolling-mill pliers, hammers, diagonal pliers, armature shafts, and so forth

Through-hardening applications include:

- Oil-country tubular products
- Structural members
- Spring steel
- Chain links

In this Chapter, applications and advantages of induction methods of heat treatment for some of the parts listed above will be discussed.

---

*This Chapter is reprinted from *Heat Treating*, Vol 4, *ASM Handbook*, ASM International, 1991 (Ref 1)

Surface-Hardening Applications

Crankshafts for internal-combustion engines were probably the first parts to which induction hardening techniques were applied. Because the explosive forces of the engine must pass through the crankshaft, severe demands in terms of strength and wear resistance are placed on the steel used in manufacturing the crankshaft. These demands are ever increasing with the rising horsepower ratings of engines used the automobiles, tractors, and other vehicles.

The most stringent demands are placed on the journal and bearing surface. Journals are the parts of the rotating shaft that turn within the bearings. Before the advent of induction heating, methods such as furnace hardening, flame hardening, and liquid nitriding were used. However, each of these processes presented problems such as inadequate or nonuniform hardening and distortion. Induction hardening overcomes many of these problems. Through proper selection of frequency, power, and the particular induction process, low distortion, case hardening can be done. In one of the most common steels used for crankshafts, 1045, case hardnesses over 55 HRC are readily obtained. Other advantages of the induction process for crankshafts include:

- Only the portions that need to be hardened are heated, leaving the remainder of the crankshaft relatively soft for easy machining and balancing.
- Induction hardening results in minimum distortion and scaling of the steel. The rapid heating associated with induction heat treating is advantageous in avoiding heavy scaling in other applications as well.
- Because induction heat treating processes can be automated, an induction tempering operation immediately following the hardening treatment can be done in manufacturing cells.
- The properties of induction-hardened crankshafts have been found to be superior to those of crankshafts produced by other techniques. These properties include strength, and torsional and bending fatigue resistance.

Presently crankshafts are being made from steel forgings as well as from cast iron. In the latter case, surface hardness levels of higher than 50 HRC are easily obtainable after induction heating and air quenching. The resultant microstructure is a mixture of bainite and martensite, with 100% martensite avoided to minimize the danger of crack formation at holes and eliminating the need for chamfering and polishing in these regions. The air quench allows the initial formation of bainite during cooling. After a prescribed period of time, the air quench is followed by a water quench during which the martensite phase is produced from the remaining austenite. Sufficient residual heat is left in the part to self temper the martensite.

Axle shafts used in cars, trucks, and farm vehicles are, with few exceptions, surface hardened by induction. Although in some axles a portion of the hardened surface is used as a bearing, the primary purpose of induction hardening is to put the surface under a state of compressive residual stress. By this means, the bending and torsional fatigue life of an axle may be increased by as much as 200% over that for parts conventionally heat treated (Fig. 10.1). Induction hardened axles consist of a hard, high-strength, and tough outer case with good torsional strength and a tough, ductile core. Many axles also have a region in which the case depth is kept very shallow so that the part can be readily straightened following heat treatment. In addition to substantially improving strength, induction hardening is also very cost-effective. This is because most shafts are made in inexpensive, unalloyed medium-carbon steel that is surface hardened to case depths of 2.5 to 8 mm (0.10 to 0.30 in.), depending on the cross-sectional size. As with crankshafts, typical hardness (after tempering) is around 50 HRC. Such hard, deep cases improve yield strength considerably as well.

Modern transmission shafts—particularly those for cars with automatic transmissions—are required to have excellent bending and torsional strength, as well as surface hardness for wear resistance. Under well-controlled conditions, induction hardening processes are most able to satisfy these needs, as shown by the data in Fig. 10.2, which compares the fatigue resistance of through-hardened, case carburized, and surface induction hardened axles. The induction hardening methods employed are quite varied and include both singe-shot and scanning techniques.

Induction hardening of crankshafts, axles, and transmission shafts is becoming an increasingly automated process. Often parts are induction hardened and in-line. One such line for heat treating of automotive parts is depicted schematically in Fig. 10.3. It includes an automatic handling sys-

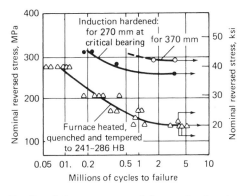

Fig. 10.1 Bending fatigue response of furnace-hardened and induction-hardened medium-carbon steel tractor axles. Shaft diameter: 70 mm (2.75 in.). Fillet radius: 1.6 mm (0.063 in.) Source: Ref 1

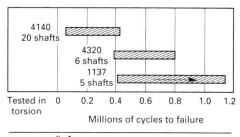

Steel	Surface hardness, HRC	Method of hardening
4140	36–42	Through-hardened
4320	40–46	Carburized to 1.0–1.3 mm (0.040–0.050 in.)
1137	42–48	Induction hardened to 3.0 mm (0.120 in.) min effective depth and 40 HRC

Fig. 10.2 Comparison of fatigue life of induction surface hardened transmission shafts with that of through-hardened and carburized shafts. Arrow in lower bar (induction-hardened shafts) indicates that one shaft had not failed after testing for the maximum number of cycles shown. Source: Ref 1

tem, programmable controls, and fiber-optic sensors. Mechanically, parts are fed by a quadruple-head, skewed-drive roller system (QHD) after being delivered to the heat-treatment area by a conveyor system. The roller drives, in conjunction with the check guides, impart both rotational and linear forward movement of the workpiece through the coil. Once a part enters the "ready position," the fiberoptic sensor senses its position and initiates the heating cycle for austenization, subsequent in-line quenching, and then induction tempering. The workpieces are round bars that are fed end-to-end continuously.

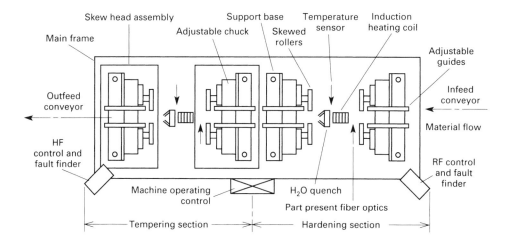

Fig. 10.3 Automated, quadruple-head, skewed drive roller system used for in-line induction hardening and tempering of automotive parts. RF, radio frequency; HF, high frequency. Source: Ref 1

In the hardening cycle of the QHD system, the induction power supply frequency is general either in the radio frequency range (approximately 500 kHz) for shallow cases or in the range for 3 to 10 kHz if deeper cases are needed. For rejection purposes, a temperature monitor senses if the workpiece has been either under-heated or overheated. Assuming that the workpiece has been heated properly, it then passes through a quench ring. After quenching, the workpiece is moved into the induction-tempering part of the heat treating line. Again, a fiber-optic sensor senses the presence of the workpiece and begins the heating cycle, generally using a lower frequency power supply (lower frequency can be used because the workpiece is still magnetic during tempering and accordingly has a shallower reference depth). Depending on the surface hardness as-quenched and the desired final hardness, the desired tempering temperature can be as high as approximately 400 °C (750 °F). As discussed in the Chapter "Tempering" in this book, induction tempering requires a higher tempering temperature than furnace because of the short heat cycle. When the tempering is complete, the workpiece is moved onto a conveyor for transportation to grinding.

The control system of this line is designed to allow decision making by a programmable controller. Thus, all aspects of the heat treating process and mechanical operations are preprogrammed and may be changed easily to accommodate different part sizes and heat treating parameters. With such a process, users have been able to increase production rates more than threefold over those obtainable with conventional heat treating lines.

Gears. Reliability and high dimensional accuracy (to ensure good fit) are among the requirements for gears. Keeping distortion as low as possible during heat treatment is very important. Induction heat treating is one of the very important processes used for heat treatment of gears. Gears, because of the wide varieties, sizes, and differences in tooth profiles, represent unique applications. External spur and helical gears, bevel and worm gears, internal gears, racks, and sprockets are good examples of the kinds of gears of which the size can range from less than 6 mm (0.25 in.) to greater than 3 m (12 feet). As with shafts, the hardened pattern may be through the cross section, as with small armature shafts, to single-teeth case hardening, as is done with large gears. A wide variety of frequencies and induction processes are used, because of the way the induced currents are produced in gear teeth with different profiles, sizes, and pitches. The heat treating processes use single-shot heating techniques and a variety of scanning techniques. A wide number of different frequencies are used to accommodate the different patterns and tooth profiles.

The size of gear, the hardening requirements, and the production requirement influence the type of induction-hardening process used. High-quantity production lots can be induction hardened single shot, whereas small quantities of large gears need to be run one tooth at a time to keep the capital equipment costs low.

170 / Practical Induction Heat Treating

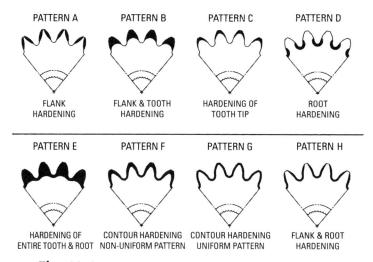

Fig. 10.4 Induction-hardening patterns for gears. Source: Ref 2

Single shot, through hardening of the ends of small armature shafts has been done since the 1950s. In addition, there is a wide variety of different types of case patterns that are produced on gear teeth. Figure 10.4 shows eight different induction patterns that can be produced with induction. Patterns A, B, and C are similar in that a portion of the tooth is hardened, but not the root. These patterns were originally used on gears with large pitch teeth. Pattern A used single-shot, channel-type coils heating the entire tooth at one time or scanning. If there is no maximum case depth specified, small gears may be through-hardened. Use of a frequency high enough that root penetration does not occur, produces patterns B and C. Figure 10.5 shows how high frequencies tend to heat the tips of teeth, while low frequencies tend to heat the root. Pattern D in Fig. 10.4 shows the root heating effect of a frequency that is too low. Patterns like this are

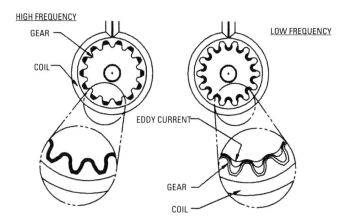

Fig. 10.5 Frequency influence on hardness profile with an encircling induction coil. Source: Ref 2

not acceptable because the upper portion of the tooth is not hard and will be subject to wear.

Gear manufacturers have found that the greatest stress on a gear is from the pitch diameter through the fillet of the root. Failure is most likely to occur at these points. Therefore, it is highly desirable that the wear surface and the root of gearing be hard. Patterns E, F, G, and H show patterns that meet these criteria. Pattern E represents one of the most common patterns produced by induction, and is produced by either single-shot heating or scanning. The specifications commonly call for the gear to be induction hardened to a minimum hardness below the root. Figure 10.6 shows the frequencies versus gear pitch that are used to produce this type of pattern. A frequency of 450 kHz has difficulty in producing case depths below 1.5 mm (0.060 in.) on even the fine-pitch gears. Single-shot hardening is limited by the power available on the power supply. Larger gears can be scanned to keep the power requirements reasonable. When distortion is excessive, the pattern requirements may be changed to that shown in Patterns F and G. With pattern F, the frequency is lowered, and the power density is increased. The case depth at the root is 30 to 40% of the case depth at the root. Pattern F attempts to produce a near contour pattern, while Pattern G attempts to produce a uniform contour. Pattern F uses pulsed or dual-frequency heating techniques. These patterns attempt to produce gears that not only quench to net shape without distortion, but also have the surface in compression so that the overall physical properties are increased.

The single power supply pulsing process uses a pre-heat of low or moderate power followed by a high power, for a very short pulse that can be 0.2 to 0.5 s. Experimentation may be needed to select the optimum frequency for a given pitch gear. The pre-heat performs two functions. First, the tip of the gear is preheated, producing an increase in the temperature at

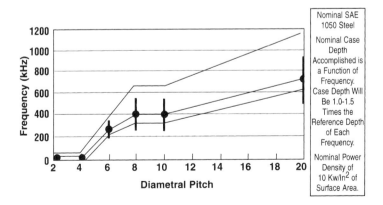

Fig. 10.6 Proper frequency selection is needed to accomplish even heating. Too low a frequency will result in field cancellation and inadequate heating; too high a frequency could overheat the surface. SAE, Society of Automotive Engineers. Source: Ref 3

the tip of the tooth so that a lower temperature rise is needed to complete austenization. Second, the preheat produces heat in the core that after quenching and cooling, produces compressive residual stress on the surface of the case. The dual frequency process uses a medium frequency to produce the pre-heat, followed by an extremely fast, high frequency austenitizing cycle. The frequencies and heating times are optimized to produce a contoured case. The dual frequency process was developed to enable induction heat treating to replace furnace carburizing.

There are several types of dual frequency systems. Figure 10.7 shows a dual-frequency system using one induction coil. This system uses the delay between the pre-heat and the fast heat cycle to transfer the coil's electrical connections from the medium frequency power supply to the RF power supply. The static heating cycle for a gear run by this system includes (Ref 4):

- Load position
- Scan up to preheat position
- Preheat 3 s
- Delay to RF 3.5 s
- RF heat 0.16 s
- Quench on, no delay (polymer based quench media)
- Quench time 8 s
- Return to load position

Figure 10.8 shows a dual-frequency set up where each of the two power supplies has its own induction coil. The gear preheated with low fre-

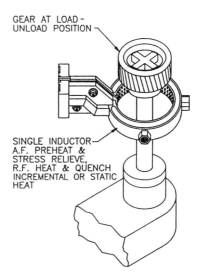

Fig. 10.7 Dual-frequency contour, 1 coil. Single induction coil provides for heating and quenching, as employed for the thermal treatments indicated. RF, radio frequency; AF, audio frequency. Source: Ref 4

Applications of Induction Heat Treatment / 173

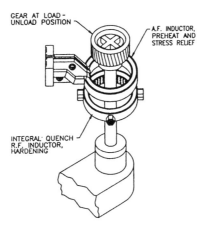

Fig. 10.8 Dual-frequency contour, two-power supplies with coils. Multi-frequency induction-hardening machine with coils designed for static heating of gear. RF, radio frequency; AF, audio frequency. Source: Ref 4

quency is then dropped to the high-frequency coil, and finally austenitized and quenched in the radio frequency (RF) coil. Figure 10.9 shows a hypothetical setup in which the gear is scanned in both the pre-heat position and the austenitizing position. Scanning permits the use of lower power densities. The pre-heat is accomplished by scanning through the coil of a medium frequency power supply. The scanning permits the power supply to have a lower output power rating than that which would be required for static heating. After the pre-heat scan, the gear is lowered into the high-heat coil of the RF power supply. Table 10.1 shows a comparison of the pre-heat scanning versus pre-heat static heating. The limitations on this process are based mainly on the size of the gear to be hardened. If the gear has a small face width, 20 mm (½ inch) or smaller, the static approach

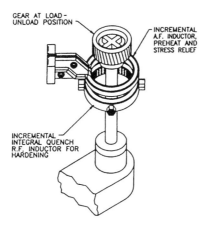

Fig. 10.9 Dual-frequency contour, two-power S, scan. Multi-frequency induction hardening machine with coils designed for scan (incremental) heating of gear. RF, radio frequency; AF, audio frequency. Source: Ref 4

may be desired. Note that the power requirements are substantial for large gears. The high power requirements may limit the ability of a user to install and use this type of process. Also, with larger gears and where a perfect contour is not desired, gears can be scanned for austenitizing in the single coil, dual-frequency configuration.

In all single-shot applications and where the overall diameter of the gear can be scanned, the gear is rotated. During the quenching of large tooth gears, the rotation speed may need to be slowed so that the gear teeth do not spin out the quenchant through a paddle-like effect. Substantial flow must be used during oil quenching to eliminate any fire hazard. Polymer quenches are widely used with quench rings and spray quenching in concentrations appropriate for the material, required hardness, and geometry of the gear. Contour-hardened gears may need very rapid and accurate application of the quenchant to force immediate and rapid quenching.

The power requirements are dictated by the power required for the application. Small, through heating applications such as the small armature shafts can be run on RF power supplies as low as 5 kW. The large, contour-hardening power supplies may require powers into the 500–1000 kW region.

Sprockets generally require heating patterns similar to pattern E in Fig. 10.4. Because of the tooth profiles, even small-diameter sprockets produce the most uniform heating, with frequencies of 25 to 50 kHz. Sprockets are hardened single-shot up to 381 to 508 mm (15 to 20 in.) in diameter, with the power supply frequency decreasing as the diameter increases.

Table 10.1 Incremental (scan) versus static contour gear hardening by dual-frequency induction heating

Sample gear development data

Gear dimensions
Major diameter 12.676 in.
Root diameter 12.00 in.
Face width 2.800 in.
Diametral pitch 8
Surface area 112 in.^{2}
Material 4150

Process requirements
Surface hardness 58–62 RC after 300 °F temper
Case depth 0.030 in. Minimum in root
Contour pattern. No more than $\frac{2}{3}$ of tooth height

Phase I Power requirements
Low frequency 1176 kW
High frequency 784 kW

Phase II Power requirements
Low frequency 480 kW
High frequency 170 kW

Source: Ref 4

The tendency is for root heating, followed by conduction of heat into the tip. Dual-frequency heating techniques have been used in which the power supply first heats the root at 3 kHz, turns off for a couple of seconds so that the power supply's heat station can switch capacitance, then turns on again at 10 kHz to top off the heat at the tooth tip. Large sprockets, such as those greater than 635 mm (25 in.) in diameter, are generally hardened by single-tooth techniques.

Powdered metal gears are usually statically induction hardened with the same types of coils as would be used on steel gears. The powdered metals heat better with high frequency, and care needs to be taken that the gears are not overheated because more porosity is produced.

Shafts, as shown in Fig. 10.10, can require that the induction-hardened pattern pass over a larger diameter, around the corner, and extend around the fillet into the smaller diameter. The basic frequency used will depend upon the material and the required case depth. The ideal situation is one in which the change in diameters is not greater than the frequency used. Because of the potential for overheating at the corner between the two diameters and the necessity to harden the fillet, the lowest possible frequency should be selected to provide deep heating. When the hardening pattern starts at the edge of the larger part, a lower power heating dwell can be used to bring the cross section up to heat, then start the scanning cycle over the smaller diameter. Another technique is to use a longitudinal coil that is designed to produce the necessary power density to scan the parts. Scanning speeds and output power may need to be changed as the part moves from the larger diameter to the smaller diameter. Finally, there is substantial application of horizontal scanners to full-length induction hardening of full-length bars and for ball screws.

Miscellaneous Applications. There are many other applications of induction surface hardening. These include uses in the ordnance, hand-tool, and automotive fields.

In the ordnance area, induction heating has been used for both surface hardening and through-hardening of armor-piercing projectiles. The induction process allows a very uniform microstructure to be obtained. It was found that induction produces a more uniform microstructure than that produced by furnace heat treating on some parts.

Induction heating has also been used for selective through-hardening and surface hardening of heads for tools such as hammers, axes, picks, and sledges. These tools are usually made of 1078/1090 steel. Lead baths were once used for the heat treating of such parts, but environmental regulations have almost eliminated the use of lead baths. Figure 10.11 shows an installation for the induction hardening of hammers. The claws are being austenitized, two at a time, using a 50 kW, 50 kHz power supply.

The split design of the claws requires that a unique coil design be used to produce uniform heating. After austenitizing, the heads are moved downwards into quench pads for polymer quenching. The faces of the striking end are austenitized on another power supply operating at 10 kHz.

Figure 10.12 shows a camshaft lob that is being induction hardened with high intensity power. Using 25 kHz, instead of 10 kHz the heat cycle was 1.2 s. The shallower case depth increased the production rate, produced straighter camshafts, and improved the lobe crowning. Vertical scanners are used to move and to position the cams in the coil on a programmed

(a)

(b)

Fig. 10.10 Shaft with changes in diameter. (a) A single-shot coil vertically heats the output shaft in eight seconds. (b) Cutaway view of single-shot shaft showing contoured pattern. Source: Ref 5

Fig. 10.11 Hammer head hardening. Hammer claws being heated to hardening temperature in a Lepel, two-position load coil and quench facility. Covers have been removed from machine. Source: Ref 6

basis. Other automotive applications include torsion bars, wheel spindles, wheel hubs, and tulips for front end drives.

Through-Hardening Applications

Through hardening and tempering are commonly done on a variety of parts through the use of induction. Through-hardening occurs either when

Fig. 10.12 Camshaft lobe. High intensity induction heating one lobe at a time (7 kW/cm^2 or 45 kW/in.^{2}). Source: Ref 5

Fig. 10.13 Diagonal pliers. Source: Ref 7

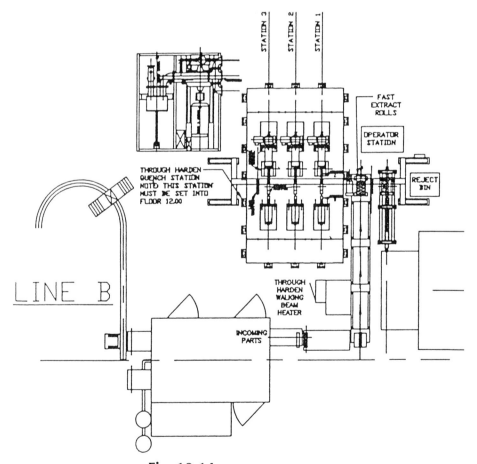

Fig. 10.14 Track links. Source: Ref 8

a portion of a part is through-heated and hardened, or when the entire part is induction hardened. Hairpin coils are used to through-harden the gripping teeth and cutting edges of many tools and cutters, with RF being very effectively used. Although the final effect appears to be a case such as the cutting edge of diagonal pliers (such as shown in Fig. 10.13), hairpin coils through heat in the area of the cutting edge only. Many parts are through heated on one wear area or cutting edge, while the balance of the part is not affected.

Figure 10.14 shows an induction heating application in which track pins are through hardened, furnace tempered, induction case hardened, and furnace tempered in one line. Two hundred links per hour for sizes ranging from 41 to 65 mm (1.6 to 2.6 in.) in diameter ranging in length from 224 to 345 mm (9 to 13.8 in.) are horizontally moved through induction hardening, quenching, and induction tempering. The pins are progressively moved by a walking beam mechanism through three different heating zones to provide a uniform austenitizing temperature. After exiting from the last austenitizing coil, the pins are conveyed to a location for pickup by a dual-grip, overhead gantry. The gantry picks up and carries the hot pins to one of three quench spindles. The dual grip enables the previously quenched pin to be removed from centers with the hot pin then placed into the centers. The pin is rotated and quenched uniformly with a polymer quenchant. During quenching the gantry moves back to pick up the next hot pin and deposits the quenched pin on a ramp for acceptance

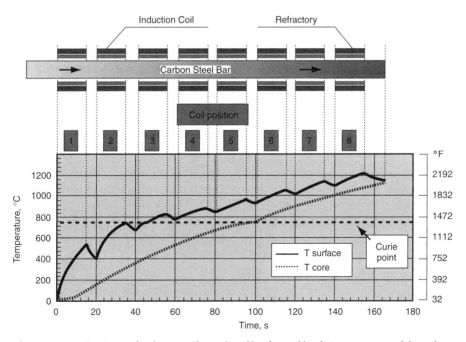

Fig. 10.15 Continuous bar heating. Thermal profile of a steel bar being as processed through an eight-coil induction line. Source: Ref 9

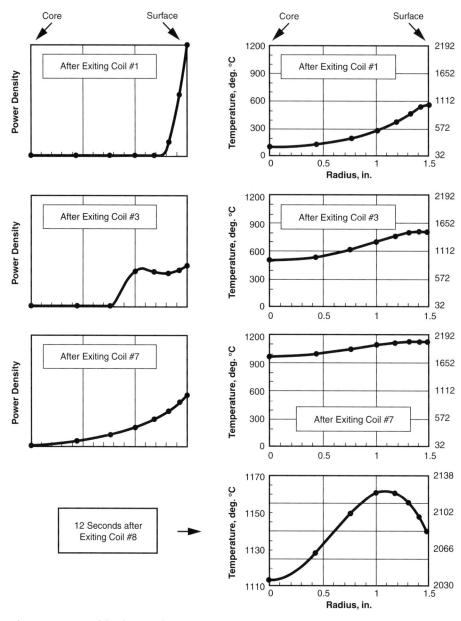

Fig. 10.16 Models of power density on continuous bar heating. Power density and temperature profiles of a bar at different positions in an in-line induction heater. Source: Ref 9

into the tempering furnace. After leaving the tempering furnace, the pins are conveyed through a cool-down chamber and located into position for pickup by a second overhead gantry. This gantry places the pins into one of three spindles for static induction heating for case hardening, with the austenitizing in the coil and quenching outside of the coil. After leaving the tempering furnace, the pins are cooled and conveyed to a grinding operation.

Modern techniques for producing long products such as cylindrical bars and rods combine the former three stages of production: casting, reheating, and rolling into a continuous line. Reheating is used to provide the bar/rod at the rolling stage with the desired temperature profile across the thickness, diameter, and length. In some cases the initial temperature of the bar/rod is uniform; in other cases the initial temperature is nonuniform due to uneven cooling as the material progresses from the caster. In recent years the bar/rod producers are selecting induction heating in the place of gas-fired furnaces. The large reduction in floor space required by induction is significant, and induction improves the bar-surface quality.

Depending on the requirements for throughput and bar size, the induction systems may consist of one or several induction coils as shown in Fig. 10.15. Computer modeling can be used to analyze the complex technological problems, so that not only is the surface-to-core temperature uniform, but also the leading and trailing bar ends are not overheated. Figure 10.16 shows the models of power density and temperature profiles of a bar at different positions in an in-line induction heating system.

REFERENCES

1. *Heat Treating*, Vol 4, *ASM Handbook*, ASM International, 1991
2. V. Rudnev et al., Gear Heat Treating by Induction, *Gear Technol.*, March/April 2000
3. Thermal Processing Databook, *Ind. Heat.*, Dec 2000
4. S.B. Lasday, Automated Multi-Frequency Multi-Cycle Induction Heat Treating of Gears with New Facility, *Ind. Heat.*, May 1991
5. M.F. Wiebowski, Induction Hardening by Design, *Met. Heat Treat.*, Jan/Feb 1995
6. A. Pers, Induction Heat Selective Hardening of 50 kHz, reprint *Met. Prog.*, April 1980
7. S.L. Semiatin and D.E. Stutz, *Induction Heat Treating of Steel*, American Society for Metals, 1986
8. S.B. Lasday, Automated Computerized Heat Treating Lines for Track Pins Beginning Operation at Caterpillar Plant, *Ind. Heat.*, Aug 1987
9. V. Rudnev et al., Efficiency and Temperature Consideration in Induction Re-Heating of Bar, Rod, and Slab, *Ind. Heat.*, June 2000

CHAPTER **11**

Induction Heat Treating Process Analysis

IN A FORMALIZED quality system, a clear understanding of the requirements for the part to be induction hardened is not only necessary but is even required under preproduction planning. This Chapter provides application and processing analysis for the induction hardening of a new part, or the review of an old part. The complete specifications for processing should be clearly stated either on the print or in a specification on an accompanying document that is referenced on the print. The material and any previous heat treatments need to be known. If a part is being designed, the factors that promote good design of a part can be considered. Potential critical stress points such as keyways, cross holes, and any nonuniform cross sections need to be identified. If the part is already designed, then these stress points need to be reviewed to see what effect they may have on successfully induction hardening a part to print specifications. Figure 11.1 provides a step by step guide with the most important factors reviewed first so that the application can be qualified before completion of the analysis. By first qualifying the application, it can be determined if the application can be done by induction and if any changes or clarifications are needed. Please note that after the initial qualification, as the analysis proceeds some of the choices, such as frequency selection and power selection, may be changed or modified because of economic factors and production requirements.

Process Qualification

Material and Prior Heat Treatment. The selection of the wrong material, or a material with a prior microstructure that is not recommended, can prevent the induction hardening process from producing good parts.

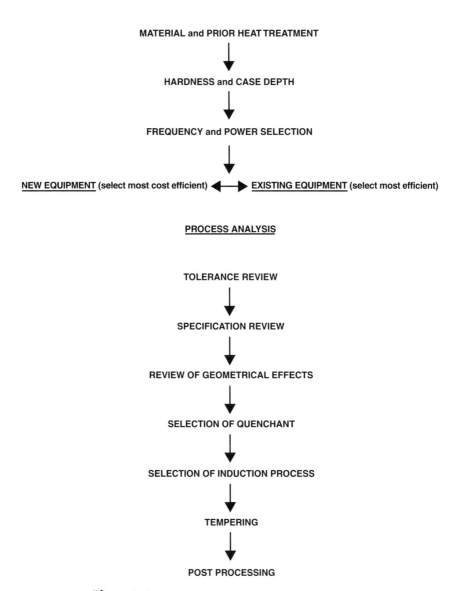

Fig. 11.1 Process qualification and analysis. Source: Ref 1

Therefore, the first step is to determine the material and any prior heat treatment from the workpiece print. This print should define the material itself, such as noting the material is *American Iron and Steel Institute* (AISI) 1045. Reference 2 has tables in the appendix that cross-reference to different standards and numbering systems. Next, review the manufacturing source of the material. The workpiece could be made from cold-rolled stock, hot-rolled stock, forged, or cast. Finally, if noted on the print, determine any prior heat treatment. This includes thermal processes such as annealing, normalizing, quench and tempering, or carbon restoring.

Hardness and Case Depth. The next step in analysis is to determine whether the specified hardness can be produced. Figure 11.2 shows the relationship between carbon content and the hardness range that can be produced. The carbon content determines whether the workpiece is capable of being quenched within the hardness ranges shown. A common error in material selection is the use of low carbon steel, unless carburized, or a lower than needed carbon content. If there is any doubt about the effect of the carbon content for a particular alloy, the J1 hardenability hardness (at or 1.5 mm, $1/16$ in.) is a good indication of the surface hardness to which an alloy with a good prior microstructure will quench. The Chapter "Heat Treating Basics" in this book discusses in detail that the best prior microstructure for induction hardening is the microstructure produced by quenching and tempering. AISI 4150 steel has high enough carbon to quench optimally to a surface hardness of 63 HRC. However, if 4150 is fully annealed, it may not quench over 54 HRC when induction hardened.

After the review determining if the part will quench to or above the desired hardness, the case-depth specifications must be reviewed. The part must have material that is capable of quenching over the desired minimum case depth. To determine this, review the hardenability chart for the material. Typical hardenability charts for some of the steels commonly induction hardened are in the Appendix "Hardenability Curves" in this book. Also, the *ASM Handbooks* may be used to find the charts for other steels not listed in the previously mentioned Appendix. Keep in mind that hardenability charts show the expected hardenability for a broad number of heats. Because of purchasing leverage, some manufacturers may be able to purchase better grades of steel than those obtainable through warehouses. However, the assumption can be made that if the specified case

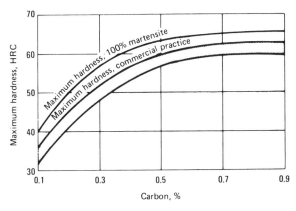

Fig. 11.2 As-quenched hardness versus carbon content. Source: Ref 2

depth, such as defined by HRC 50 or other hardness value, meets the specified depth for all heats, the workpiece will quench to the specified case depth with the correct prior microstructure.

At this point, had the qualification shown that the part and material could not be induction hardened to the specification, the analysis could proceed with three different options. The first option is to stop the analysis until a change in material or heat treatment is made. The second option is to list the specifications that the intended material is expected to produce. The third option is to specify the material and/or prior heat treatment needed and to proceed with the qualification noting this assumption in writing. An example of this would be to specify that the carbon content of the material needs to be a certain minimum value. If an AISI 1040 steel is specified with the requirement that the surface hardness is to be 61 HRC, the use of 1050 steel will readily produce this hardness. Finally, when the conclusion has been made that the material is capable of being induction hardened to the given specifications, the first steps in qualification are complete. Further discussion will be given to hardness and case-depth specifications later in this chapter. The next step in qualification is to make the initial power supply frequency and power requirements.

Frequency selection (Ref 3) is the first parameter considered for induction heating. Primary considerations in the selection of frequency are depth of heating efficiency, type of heat treatment (such as surface hardening versus through hardening), and the size and geometry of the part.

The frequency and power supplies commonly used in the induction hardening of steel are compared in Table 11.1. As shown in this tabula-

Table 11.1 Frequency selection

Case depth		Diameter		Frequency				
mm	in.	mm	in.	1 kHz	3 kHz	10 kHz	50 kHz	450 kHz
0.38–1.27	0.015–0.050	6.35–25.4	1/4–1	...	...	...	...	Good
1.29–2.54	0.051–0.100	11.11–15.88	7/16–5/8	...	...	Fair	Good	Good
		15.88–25.4	5/8–1	...	...	Good	Good	Good
		25.4–50.8	1–2	...	Fair	Good	Good	Good
		>50.8	>2	Fair	Good	Good	Good	Poor
2.56–5.08	0.101–0.200	19.05–50.8	1–2	...	Fair	Good	Good	Poor
		50.8–101.6	2–4	Fair	Good	Good	Good	Poor
		>101.6	>4	Good	Good	Fair	Good	Poor
5.08–10.0	0.200–0.400	>8	>2	Good	Good	Fair	Poor	Poor
Through hardening								
		1.59–6.35	1/16–1	...	...	...	...	Good
		6.35–12.7	1/4–1/2	...	...	Fair	Fair	Good
		12.7–25.4	1/2–1	...	Fair	Good	Good	Fair
		25.4–50.8	1–2	Fair	Good	Fair	Poor	...
		50-8–76.2	2–3	Good	Good	Poor	...	...
		76.2–152.4	3–6	Good	Poor	Poor	...	...
		>152.4	>6	Poor	Poor	Poor	...	...

Good indicates most efficient frequency. Fair indicates the frequency is less efficient. Poor indicates not a good frequency for this depth. The coil power density must be kept within the recommended ranges. Source: Ref 1

tion, the lower frequencies are more suitable as the size of the part and the case depth increases. However, because power density and heating time also have an important influence on the depth to which the part is heated, wide deviations from Table 11.1 may be made with successful results. This interrelationship is shown in Fig. 11.3 in terms of case depth, frequency, and power density for surface-hardened steel. In some instances, the determining factor in selecting the frequency is the power required to provide power density sufficient for successful hardening, as lower frequency induction equipment is available with higher power ratings and lower cost.

The equation given for reference depth, d, in the Chapter "Theory of Heating by Induction" in this book can be used to estimate the optimal power supply frequency for induction hardening of steel. For surface hardening, the desired case depth is typically taken to be equal to about one-half the reference depth required for austenitizing for selection of minimum frequency. Table 11.1 shows the practical effect of frequency selection for hardening to various case depths. By contrast, when through hardening is desired, the frequency is usually chosen such that the reference depth is a fraction of the bar radius (or an equivalent dimension for parts that are not round). This is necessary in order to maintain adequate "skin effect" and to enable induction to take place at all. If the reference depth is chosen to be comparable to or larger than the bar radius, there will be two sets of eddy currents near the center of the bar induced from diametrically opposed surfaces of the bar. From a practical viewpoint, it is as if the induced currents cancel each other. To avoid this, frequencies for through hardening are often chosen so that the reference depth does not exceed approximately one-half the thickness for plates and slabs when using solenoid coils. When the bar diameter is less than four reference

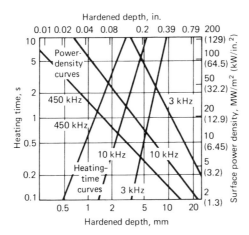

Fig. 11.3 Interrelationship among heating time, surface power density, and hardened depth for various induction generator frequencies.
Source: Ref 3

depths, or slab thickness is less than two reference depths, the electrical efficiency drops sharply. By contrast, little increase in efficiency is obtained when the bar diameter or slab thickness is many times more than the reference depth.

Typical frequency selections for induction hardening of steel workpieces are listed in Table 11.1 and Fig. 11.4. Those for surface hardening will be examined first. For very thin cases such as 0.40 to 1.25 mm (0.015

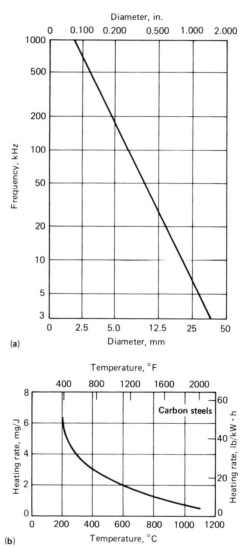

Fig. 11.4 Typical frequency selections and heating rates for induction hardening of steel parts. (a) Relationship between diameter of round steel bars and minimum generator frequency for efficient austenitizing, using induction heating. (b) Heating rate for through heating of carbon steels by induction. For converted frequencies, the total power transmitted by the induction to the work is less than the power input to the machine because of converter losses. See also Fig. 11.1. Source: Ref 3

to 0.050 in.) on small diameter bars, which are easily quenched to martensite, relatively high frequencies are best. If the reference depth is equated to the case depth, the best frequency for a 0.75 mm (0.030 in.) deep case on a 13 mm (0.05 in.) diameter bar is found to be around 450 kHz. When the surface of a larger diameter bar is hardened, particularly when the case is to be deep, the frequency is often chosen so that the reference depth is several times the desired case depth. This is because the large amount of metal below the surface layer to be hardened represents a large thermal mass which draws heat from the surface. Unless very high power densities are employed, it is difficult to heat only the required depth totally to the austenitizing temperature. As an example, consider the recommended frequency for imparting a 3.8 mm (0.15 in.) hardened case to a bar 75 mm (3 in.) in diameter. If reference depth were equated to the case depth and a frequency of about 10 kHz were selected, only "fair" results would be produced. 3 kHz, with its deeper reference depth, is more efficient at producing the deeper case depth on large diameters and is accordingly rated at "good."

For through hardening of a steel bar or section, the optimum frequency is often based on producing a reference depth about one-fourth of the bar diameter or section size. For instance, through heating and through hardening a 64 mm (2.5 in.) diameter bar would entail using a power supply with a frequency of about 1 kHz. If lower frequencies were employed, inadequate skin effect (current cancellation) and lower efficiency would result. On the other hand, higher frequencies might be used. In these cases, however, the power supply output would have to be low enough to allow conduction of heat from the outer regions of the steel part to the inner ones. Otherwise, the surface may be overheated, leading to grain growth or even melting.

Power Density and Heating Time. Once the frequency has been selected, varying the power density and heating time can produce wide ranges of temperature profiles. Selection of these two heating parameters depends on the inherent heat losses of the workpiece (from either radiation or convection losses) and the desired heat conduction patterns of a particular application.

In through heating applications, the power needed is generally based on the amount of material that is processed per unit time, the peak temperature, and the material's heat capacity at this temperature. Power specification for other operations, such as surface hardening of steel, is not as simple because of the effects of starting material condition (prior microstructure) and the desired case depth.

Surface heating is used primarily in the surface hardening of steel parts such as shafts and gears. In this type of application, high power densities and short heating times are used when thin case depths are desired.

Typical power density recommendations for surface hardening of steel are given in Table 11.2. These are based on the need to heat very rapidly to the austenitizing temperatures shown in Table 11.3 and have proven through years of experience to be appropriate. When using these or other fixed ratings, however, the effect of heating time and frequency on case depth (Fig. 11.3) must be considered. Higher frequencies and higher power densities produce smaller case depths. Thus, lower frequencies at higher power densities can be used to produce the minimum possible case depth for that frequency.

Through Heating. To allow time for the heat to be conducted to the center of the workpiece, power ratings for through-hardening of steel are much lower than those for surface hardening. After a while, the rate of increase of the surface and center temperatures becomes comparable due to conduction, and a fixed temperature differential persists during further heating. On larger workpieces the entire cross section can be austenitized, but temperature differential will always be present from surface to core center. Table 11.1 shows frequency efficiency versus diameter of round bar. In addition to the methods used by Tudbury (Ref. 4), the allowable temperature differential permits the power supply output ratings to be selected. The process involves selecting the frequency and calculation of the bar diameter (or section size) to reference depth, a/d. For most through heating applications this ratio will vary from around four to six. Then the values of the thermal conductivity are used to estimate the induction thermal factor. Finally, the total power required to heat the bar is calculated.

In addition to these estimates, if total accuracy is needed radiation heat loss must also be considered when calculating power requirements. At 875 °C (1600 °F) the losses are 0.010 kW/cm² (0.065 kW/in.²) of surface. The losses double when heating to 1100 °C (2000 °F). On the continuous heating of large surfaces, these losses should be added to the power requirements.

Table 11.2 Power density required for surface hardening

Frequency, kHz	Depth of hardening(a), mm	in.	Input W/mm² Low (b)	Optimum (c)	High (d)	Input kW/in.²(e) Low (b)	Optimum (c)	High (d)
500	0.38–1.14	0.015–0.045	10.9	15.5	18.6	7	10	12
	1.14–2.29	0.045–0.090	4.7	7.8	12.4	3	5	8
10	1.52–2.29	0.060–0.090	12.4	15.5	24.8	8	10	16
	2.29–3.05	0.090–0.120	7.8	15.5	23.3	5	10	15
	3.05–4.06	0.120–0.160	7.8	15.5	21.7	5	10	14
3	2.29–3.05	0.090–0.120	15.5	23.3	26.35	10	15	17
	3.05–4.06	0.120–0.160	7.8	21.7	24.8	5	14	16
	4.06–5.08	0.160–0.200	7.8	15.5	21.7	5	10	14
1	5.08–7.11	0.200–0.280	7.8	15.5	18.6	5	10	12
	7.11–9.14	0.280–0.360	7.8	15.5	18.6	5	10	12

Note: This table is based on use of proper frequency and normal over-all operating efficiency of equipment. Values in table may be used for static and progressive methods of heating; however, for some applications, higher inputs can be used when hardening progressively. (a) For greater depth of hardening, a lower kilowatt input is used. (b) Low kilowatt input may be used when generator capacity is limited. These kilowatt values may be used to calculate largest part hardened (single-shot method) with a given generator. (c) For best metallurgical results. (d) For higher production when generator capacity is available. (e) Kilowattage is read as maximum during heat cycle. Source: Ref 3

Table 11.3 Approximate induction austenitizing temperature

Steel	Carbon, %	Austenitizing temperature °C	°F
1022	0.18/0.23	900	1650
1030	0.28/0.34	875	1600
10B35	0.32/0.38	855	1575
1040	0.37/0.44	855	1575
1045	0.43/0.50	845	1550
1050	0.48/0.55	845	1550
1141	0.37/0.45	845	1550
1144	0.40/0.48	845	1550
1541	0.36/0.44	845	1550
Stress proof	0.36/0.44	845	1550
Fatigue proof	0.36/0.44	845	1550
4130	0.28/0.33	870	1600
4140	0.38/0.43	875	1600
4150	0.48/0.53	845	1550
ETD 150	0.48/0.53	845	1550
4340	0.38/0.43	845	1550
5160	0.56/0.64	845	1550
52100	0.98/1.1	800	1475
8620	0.18/0.23	875	1600
1018 Carb.	0.9 nom	815	1500
1118 Carb.	0.9 nom	815	1500
8620 Carb.	0.9 nom	815	1500
5120 Carb.	0.9 nom	815	1500
416 SS	<0.15	1065	1950
420 SS	>0.15	1065	1950
440C SS	0.95/1.2	1065	1950
O1	0.9	815	1500
D2	1.5	1020	1875
D3	2.25	980	1800
A1	1	980	1800
S1	0.5	955	1750

The induction austenitizing temperature can be up to 200 °F (110 °C) higher depending upon the prior microstructure and the rate of heating. Source: Ref 1

However, through the use of tables and knowing the effective weight of steel to be heated per hour, it is easy to determine the power supply requirements. From knowing the effective heat cycle time, the effective weight of steel to be heated per hour can be calculated. If the process is not continuous, the effective heat cycle can be calculated by knowing the operating efficiency and number of parts to be heated per hour. If the required heating time is 10 s, then divide one hour (3600 s) by this heat cycle of 10 s to get an effective heating rate per hour of 360 parts. Then multiply by the effective weight to be heated to obtain the pounds per hour. If the 360 parts have an effective weight of 600 lbs and need to be heated to 875 °C (1600 °F), then 0.8 mg/J (4 lbs/kW-h) or 250 kW is needed (Fig. 11.4b). Table 11.1 can be reviewed to determine frequency ranges that can be used. Finally, the approximate power densities required for the through heating can be obtained from Table 11.4. When through heating is desired, lower power densities are used than those used for surface hardening. Because the price of power per kW of power supply decreases with frequency, sometimes there are trade-offs between the optimum frequency and the frequency selected. On large capacity, through heating installations such as the continuous hardening of bar stock, more

Table 11.4 Approximate power densities required for through-heating of steel for hardening, tempering, or forming operations

Frequency(a), Hz	Input(b)									
	150–425 °C (300–800 °F)		425–760 °C (800–1400 °F)		760–980 °C (1400–1800 °F)		980–1095 °C (1800–2000 °F)		1095–1205 °C (2000–2200 °F)	
	kW/cm^2	kW/in.^{2}	kW/cm^2	kW/in.^{2}	kW/cm^2	kW/in.^{2}	kW/cm^2	kW/in.^{2}	kW/cm^2	kW/in.^{2}
60	0.009	0.06	0.023	0.15	(c)	(c)	(c)	(c)	(c)	(c)
180	0.008	0.05	0.022	0.14	(c)	(c)	(c)	(c)	(c)	(c)
1000	0.006	0.04	0.019	0.12	0.08	0.5	0.155	1.0	0.22	1.4
3000	0.005	0.03	0.016	0.10	0.06	0.4	0.085	0.55	0.11	0.7
10 000	0.003	0.02	0.012	0.08	0.05	0.3	0.070	0.45	0.085	0.55

(a) The values in this table are based on use of proper frequency and normal overall operating efficiency of equipment. (b) In general, these power densities are for section sizes of 13 to 50 mm (½ to 2 in.). Higher inputs can be used for smaller section sizes, and lower inputs may be required for larger section sizes. (c) Not recommended for these temperatures. Source: Ref 1

than one frequency power supply may be used in line to promote the most efficient heating. Large tonnage installations may use frequencies ranging down to line frequency for preheating below the Curie Temperature (also providing reduced thermal shock), and higher frequency are used above the Curie. Also, lower frequencies are best for induction tempering because the deeper penetration of the induced current during heating requires less heat conduction to produce a uniform temperature.

As shown, frequency and power selection can have two objectives. First, if the purpose of the analysis is for the selection of new equipment, the lowest cost frequency will need to be determined. If the purpose of the analysis is to determine which equipment in house should be used, then the selection can be made from the viewpoint of the most efficient production processing cost equipment. As discussed, a shallow case depth might be more efficiently produced with higher frequency, while a deeper case depth is more efficiently produced by lower frequency.

Tolerance review is the next step of analysis. At this point of the review process, the broad qualification objectives of material review and power supply selection have been completed. Following is a review of the print of the part to be induction hardened, along with any pertinent specifications. The print should indicate the area to be induction hardened and any specifications, either listed directly on the print or referred to by a specification number.

- *Machining tolerances:* May be shown either on the print dimension lines or in one of the information boxes at the bottom of the print. If the machining tolerances are in a box, the number of decimals defines the machining tolerance. For example 0.xxx dimension might have a 0.05 mm (0.002 in.) tolerance. The machining tolerances are important because the pattern length tolerance may be indicated from a location that in itself has a machining tolerance that can affect the pattern tolerance. Induction pattern lengths should not be specified to the same tolerances as machining tolerances.

- *Overall part dimensions:* Should be clearly shown. Locations for holding the workpiece should be determined.
- *Pattern length location and tolerances:* Pattern starting locations need to be clearly indicated from a machined surface, preferably from the location at which the part will be held or located. Induction pattern locations and tolerances need to be stated separately from machining tolerances. The pattern length, depending on the frequency of the induction power supply, needs as large a tolerance as possible. One to two mm (0.040 to 0.080 in.) or greater is tight for radio frequencies (RFs), and tolerances can range up to 6.1 mm (0.25 in.) for the lower frequencies. Because tighter tolerances represent more setup and quality control cost, the pattern tolerances should be reasonably specified. Some parts require only minimum pattern length tolerances.

Specification Review. ASTM testing standards are discussed in the Chapter "Standards and Inspection" in this book as related to hardness testing, while the Chapter "Nonconforming Product and Process Problems" in this book discusses testing and inspection for induction. There are currently no national standards for the other induction heat treating specifications. This includes such items as definition of *case depth* (apparent or effective), pattern tolerances, microstructure requirements, special inspection requirements, or standards as listed in separate individual company standards. A wide number of companies have individual standards for the measurement of case depth and their own microstructural requirements. These specification numbers should be referenced somewhere on the print of the part to be induction hardened.

Case Depth Required and Definition. When no method of case depth specification is specified, the most economical method of case depth verification is through the use of an etchant to define the total (or apparent) case with measurement of the etched depth. When the case depth is measured by determining a hardness value, the method of measurement is called *effective case*. For this definition medium carbon steels most often use HRC 50, or its equivalent. There is no standard on effective case depth for induction-hardened cases, so that the definition will vary by manufacturer. Most manufacturers use a hardness value below 50 HRC when the carbon drops below 0.45%. The different types of common case depth specifications are:

- Minimum total case or effective case depth, no maximum (part can be through hardened)
- Minimum and maximum total case
- Minimum and maximum effective case
- Minimum effective case, maximum total case
- Minimum effective case at HRC 50, minimum transition case at another hardness value such as HRC 35 (this is for deep case requirements)

Geometrical Effects. The shape of the workpiece and the corresponding shape of the induction coil affect the induction heating process. The ideal shape is a round workpiece, such as a bearing. When the shape or required hardness patterns become nonuniform, the analysis needs to be more detailed because of both overheating and under heating tendencies. This next section will discuss the effects of some of the different types of workpieces and complex shapes.

Part Shape with Regard to Location of the Pattern. The easiest application of induction is where there is symmetry in respect to the workpiece and the induction pattern required. Round workpieces are ideal because the induction coil can be contoured around the diameter (or in the inside diameter, or ID of a hollow workpiece). Square and rectangular cross sections tend to have deeper heating at the corners, with the higher frequency producing more heat at the corners. Workpieces, such as diagonal pliers, easily have the cutting edge hardened. Dies that have an edge hardened on one side over a long length set up bending stresses that will tend to make the part bow, even if restrained during heating (some cutters and ways have opposing sides hardened to minimize this tendency toward bowing). Hairpin and channel coils can sometimes be contoured to follow the outline of the workpiece. Other times thin cross sections are prone to overheating. There are times when the best information on the ability to heat a given workpiece is that obtained from running test heats. The different types of coil design are discussed in the Chapter "Induction Coils" in this book.

Part Shape with Regard to Any Changes in Diameter or Cross Section in the Area to be Heated. Changes in diameter require analysis to determine whether the desired induction pattern can be produced. Difficulty occurs where the change in cross section is greater than the reference depth. Although in some circumstances a timed dwell can be used to initially soak in the heat at the junction between the two diameters, if the diameter change is not too great, conventional encircling coils tend to skip and not to heat the fillet between the two diameters. Longitudinally oriented coils may need to be used in this situation. Workpieces such as shafts, which require both a flange face and the adjacent diameter to be hardened, may require the "Z" type coils as shown under "Coil design" in this Chapter. These coils are single-shot coils, but they induce the current into the flange around the circumference while inducing the current into the shaft longitudinally.

Tapered shafts or shafts with changes in diameter greater than 1 in. are difficult to harden. The coil losses will increase to the point that not enough power is induced into the workpiece to effectively heat the smaller diameters. Another difficult example is the scan hardening of the diameter, while trying to harden the face at the end. Not enough current can be induced to heat the face without overheating the diameter adjacent to the face.

Holes. When under an induction coil, the current flow around a hole heats nonuniformly with more heat produced around the hole area. Small

Induction Heat Treating Process Analysis / 195

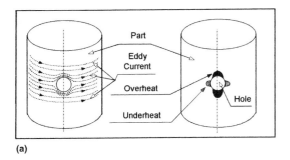
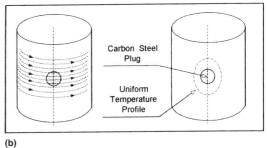
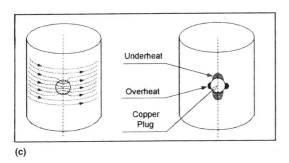
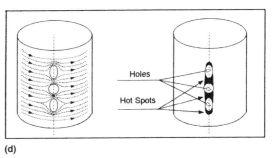
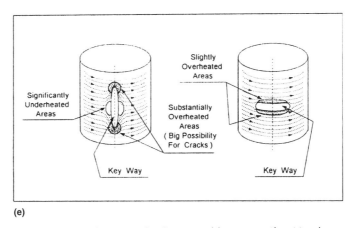

Fig. 11.5 Crossholes overheating. Eddy-current distribution and heat nonuniformities due to presence of transverse holes. (a) Transverse hole, no plug; (b) carbon steel part and carbon steel plug; (c) carbon steel part and copper plug; (d) multi-holed part, no plugs. Source: Ref 5

holes, such as 1.5 mm (0.060 in.) are not affected. As the hole size and the frequency increase, the overheating effect increases as well. The overheating produces a tendency toward quench cracking or even melting. Machining chamfer relief on the edge of a hole will reduce this overheating tendency. Figure 11.5 (a) and (d) show the overheating effect of holes in a round bar. When holes overheat, metal plugs inserted before heating will stop the current concentration and overheating. The plugs must be inserted tight enough that they do not fly out when the part is heated. If the plugs are made of soft material, they can be drilled out. Copper or brass plugs can be used, but they have a tendency toward producing nonuniform heat, as shown in Fig. 11.5(c). Low-carbon steel, because it is similar in resistivity and conductivity to carbon steel, will tend to negate the effects of the hole, as shown in Fig. 11.5(b). Because low carbon will not harden, the low-carbon plugs can be drilled out after quenching. Small pilot holes can even be put in the plugs to help with their removal. Some use has been made of non-metallic plugs, but they are difficult to use on a production basis. Finally, there have been installations in which air is blown into a longitudinal hole in the part that connects to the cross hole. The air prevents quench cracking.

- *Cross holes in the pattern:* Are sensitive to overheating as the hole size increases and frequency decreases. For instance, at 450 kHz, holes less than 1.5 mm (0.060 in.) in diameter do not generally provide any problems. As the holes size approaches 3.0 mm (0.125 in.) in diameter, the tendency towards overheating increases. At diameters greater than 3 mm (0.125 in.), the holes may need to be plugged. At 10 kHz, the critical hole size is about 6 mm (0.375 in.) in diameter. At 3 kHz the critical hole size is about 12.5 mm (0.500 in.).
- *Cross holes close to the edge of the pattern:* Can cause a localized overheating effect at the edge of the pattern, with the effect that the pattern becomes elliptical at the areas of the holes. If holes are present, the pattern tolerance needs to be larger than that required if the holes were not present.
- *Longitudinal holes close to the bottom of the pattern:* Cause distortion of the magnetic flux field because the normal heat flow into the core of the part is upset. Because there is no material in the hole to conduct heat, the area between the hole and the case will be hotter. If the hole is close enough to the austenitized area, the longitudinal area in the case will actually form a hot spot that is overheated. Cracking and distortion can result. Whenever possible, these holes should be drilled after the parts are induction hardened. If this cannot be done, it may be necessary to plug these holes. If the holes are threaded, bolts can be inserted to help with the heat flow and to reduce the overheating.
- *Longitudinal holes in the core (under the case):* Except on centerline, tend to promote distortion as discussed in the Chapter "Heat Treating Basics" in this book. Figure 11.6 shows the overheating effect of longitudinal holes. The drilling of these holes after the parts are induction hardened will promote lower distortion. If distortion is excessive and the holes are threaded, bolts can be inserted to reduce distortion.

Keyways, unless required by the specifications, should not be hardened. However, there are parts which require the strength that is needed in the area of keyways. Keyways will always promote nonuniform heating.

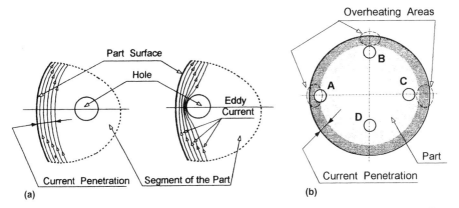

Fig. 11.6 Longitudinal holes overheating. (a) Eddy-current redistribution due to presence of longitudinal hole; (b) overheating areas due to presence of longitudinal holes. Source: Ref 5

Figure 11.5(e) shows the heating effect of keyways, with the particular tendency toward overheating in one area while underheating in another. Nonconcentric hardened cases are produced, leading to bowing or distortion. In addition, the fillet areas of the keyways are stress risers that serve to initiate quench cracks. It is usually better to use a frequency that will harden the entire keyway cross section than to use a higher frequency and not harden the bottom of the keyway. There have been other ingenious ways that the keyways have been plugged, including fastening in metal plugs and molding various materials. However, these techniques are used more as a salvage technique than for steady production. Whatever plug is used must not come out of the keyway during austenitizing and quenching.

Bores and IDs. If the part is a tube or has a bore with a thin cross section, there may be a problem with through heating unless secondary cooling is used on the outside diameter (OD) during heating. The bore must be large enough that an induction coil can fit inside the bore. While there have been some induction coils developed to heat under 12 mm (0.50 in.) diameters, larger diameters are preferred. If possible keyways in bores, even if the OD is being hardened, should always be broached after induction hardening because they are stress risers and may promote cracking during quenching. Finally, heat transfer does not occur as rapidly when heating a bore. It is easier to control shallow cases, and correspondingly is more difficult to produce deep cases.

Tubes and Sleeves. Hollow workpieces act in an induction coil differently than solid workpieces. The hollow core reacts differently with the electromagnetic flux field so that tubular cross sections tend to through heat. Lower frequencies can be used to heat hollow parts than are required for solid parts of the same diameter.

Selection of Quenchant. The selection of the quenchant is important because quenching is part of the process cycle and affects the hourly production rate. In order to calculate the hourly production rate, as outlined next, the method of quenching must be determined. Quenching a workpiece generally takes as long as or longer than hardening it. With alloy steels and slow quenchants this becomes even more pronounced. The theory of quenching was discussed in the Chapter "Quenching" in this book. Table 11.5 lists recommended quenchants for the various steels commonly induction hardened and where there is danger of quench cracking. When a particular grade of steel has a propensity toward quench cracking, the shape of the parts and the metallurgical characteristics influence quenchant selection and quenching technique. The selection of the quenchant will influence the selection of the induction process that is going to be used. Longer quenching times may require outside-the-coil quenching, multiple heating and quenching positions, and the use of secondary quenching in order to optimize production rates.

Selection of Induction Process. The number of workpieces to be run per hour and the operating efficiency help to determine the fixturing and

Table 11.5 Quench response

Steel	Carbon %	As quenched hardness (HRC)	Quenchant type, static htg.	Quenchant type, scanning	Tendency for quench cracks
1022	0.18/0.23	30/44	Water	Water	No
1030	0.28/0.34	40/52	Water	Water	No
10B35	0.32/0.38	55/58	Water	Water	No
1040	0.37/0.44	50/60	Water	Water	No
1045	0.43/0.50	54/64	Water/P	Water/P	Yes
1050	0.48/0.55	55/65	Water/P	Water/P	Yes
1141	0.37/0.45	48/56	Water/P	Water/P	No
1144	0.40/0.48	55/62	Water/P	Water/P	Yes
1541	0.36/0.44	53/60	Water/P	Water/P	No
Stress proof	0.36/0.44	55/62	Oil/P	Water/P	Yes
Fatigue proof	0.36/0.44	55/62	Oil/P	Water/P	Yes
4130	0.28/0.33	49/56	Oil (24% P)	Water/P	No
4140	0.38/0.43	53/60	Oil/P	Water/P	Yes
4150	0.48/0.53	59/64	Oil (24% P)	Water/P	Yes
ETD 150	0.48/0.53	59/64	Oil (24% P)	Water/P	Yes
4340	0.38/0.43	53/60	Oil	24% P	Yes
5160	0.56/0.64	58/65	Oil	P	Yes
52100	0.98/1.1	60/66	Oil/P	P	Yes
8620	0.18/0.23	36/46	Oil (24% P)	P	No
1018 Carb.	0.9 nom	60/66	Water	...	No
1118 Carb.	0.9 nom	60/66	Water	...	No
8620 Carb.	0.9 nom	60/66	2/20% P	...	No
5120 Carb.	0.9 nom	62/66	2/20% P	...	No
416 SS	<0.15	36/45	Oil	...	No
420 SS	>0.15	46/55	Oil	...	No
440C SS	0.95/1.2	58/62	Oil	...	No
O1	0.9	59/63	Oil	...	Yes
D2	1.5	61/64	Oil	P	Yes
D3	2.25	63/66	Oil	P	Yes
A1	1	61/65	Oil	...	Yes
S1	0.5	56/59	Oil	...	Yes
Gray cast iron		>50	P	...	No
Ductile cast iron		>50	P	...	No

P is polymer quench with the concentration depending on the part geometry and hardness requirements. Caution must be used in selection of the quenchant where there is a tendency toward quench cracking because of the part geometry and hardness requirements. 24% polymer is considered in this table to be the same rate as that produced by a fast speed oil. These recommendations are from commerical practice. Optimum hardness can be produced when the microstructure before induction hardening is optimum. Bold quench "yes" means this material is very prone toward producing quench cracks. nom, nominal. Source: Ref 1

type of induction process. From this information, hourly production rates need to be established. Low production requirements will generally require less mechanization unless there is a workpiece weight or handling problem. Discussion of the various induction-hardening techniques and processes will be discussed later in this chapter. With high production requirements, production rate calculations may need to be made with several types of production modes versus cost of equipment and operation. An example of this is the use of several scanners each hardening parts with low kW power supplies as opposed to one single-shot system using a large kW power supply. As previously discussed, the area to be induction hardened should be calculated in square inches. An examination of the available power will determine the power density and the type of processing that should be done (i.e., single-shot versus scanning). The higher productivity produced by single-shot induction hardening may be needed. Also, at this time different frequencies may be considered at different power levels to try to best meet the production requirements. Higher power, lower frequency may be used, although more inefficient, because the overall economics are superior.

Coil design was discussed in the Chapter "Induction Coils" in this book. Different designs of coils, particularly if magnetic flux concentrators are used, operate with different efficiencies. Select the type of induction coil that will best produce the desired induction pattern, keeping in mind the desired production rate. The coil selected must produce the required pattern, or no amount of unique mechanization will produce a good part. Through the use of flux concentrators, coils can be designed for specific patterns that were formerly very difficult to produce. After the coil design is selected, determine the process and fixturing that will work best. An example is a spindle (Fig. 11.7) with small production requirements that has a substantial change in diameter. This figure shows a non-encircling induction coil with flux concentrator that is used for single-shot hardening. However, a shorter non-encircling coil can be designed to be used with a scanner and a lower output power supply. This system would start with a heat dwell at the fillet and would then scan the rest of the shaft. The shorter coil uses the same power density. It should be noted that some new parts that have no processing history require development programs for the process and coil design.

Distortion may or may not be important. Some workpieces are finish machined or ground after induction hardening. One technique is to pre-establish the distortion and then to machine the workpiece with a tolerance that allows for distortion. Another technique is to establish a heating-quenching cycle that produces no distortion. This can involve pre-heating and the timed use of primary and secondary quenchants.

Fig. 11.7 Example of shaft with changes in diameter. Source: Ref 5

Holding Locations for the Workpieces. The ability to maintain process control and to produce conforming induction-hardened parts usually requires that the workpieces be positioned in exactly the same way in the induction coil for each heating cycle. Thus it is important to determine what surfaces on the workpiece can be used to hold the part. As previously discussed in the review on tolerances, the location point must have tight tolerances in reference to the area to be heated. Machined surfaces are preferred. The holding fixture or nest must not interfere with the magnetic flux field of the coil. When the workpieces are held by cupping nests, the ends must be flat and free of burrs. If the workpieces are held between centers, the center holes must be drilled to the same depth.

Automotive safety items require special, bulletproof processing. The production process must ensure that every part that is shipped is in specification. If statistical process control (SPC) is required, decisions must be made regarding the critical specifications that are to be tested and to be recorded for SPC purposes.

Tempering is discussed in the Chapter "Tempering" in this book. The requirement for tempering may be stated in the print or specifications. Where parts are austenitized and quenched to a hardness that is above the specified hardness range on the print, the parts must be tempered. If the parts quench within the specified hardness range, but tempering is optional or not listed, tempering should be considered if the parts have designs that might promote brittle failures or if the parts are ground after induction hardening. If the parts are to be furnace tempered, the number of parts per load for batch tempering or the number of parts per hour for continuous tempering should be calculated. If the parts are to be induction tempered, the induction tempering process needs to be designed.

Post-Production Processing. On some parts, straightening may be required. On all parts after processing, there may be requirements that need to be reviewed for cleaning the parts or for applying a rust preventative.

Workhandling Equipment Selection

In order to induction harden a workpiece in a controlled process, the workpiece must be located consistently in the same, exact location in the induction coil. Most workpieces are rotated when possible. Depending on process requirements, workpieces may be loaded and unloaded manually or mechanically (or a combination of both). There are many different types of mechanical fixtures that can be used to accomplish this. Low production is more likely to involve manual loading and unloading, while high production is more likely to involve mechanized loading. Furthermore, some of the fixtures move the parts through the induction coil. If the application is being analyzed for new equipment, then the process can be optimized for whatever criteria are needed. This could range from the cheapest equipment to do the job, from a capital equipment cost, to what equipment in-house can do the job.

General Types of Work-Handling Fixtures. There are two broad classifications of heating: in-place heating and scanning. In-place heating is usually named static heating or single-shot heating. The workpiece is positioned in the coil, a timed cycle is used for heating, and then the workpiece is quenched. The difference between static heating and single-shot heating is that single-shot heating involves the use of a longitudinally orientated coil with the workpiece rotated in the coil. When an encircling coil is used during static heating, the part is usually rotated. However, static heating may use hairpin or channel-type coils in which the workpiece is not rotated. Depending on the application, the coil design, and the fixturing, one or more workpieces or areas of a workpiece may be heated at a time. Most workpieces that can be heated statically can be heated single shot through the use of higher power with single-shot tooling. Scanning usually involves a timed heat dwell to start the heat cycle followed by scanning one or more workpieces through the coil, heating and quenching progressively. A variation of scanning is where the part is oscillated in the coil and quenched at the end of the heating cycle. With either classification, there are a number of different fixturing techniques and part orientations possible. For instance, parts may be scanned with either vertically or horizontally orientated fixtures.

One Position, Heating. The workpiece is manually placed in a location in the coil; the operator depresses the "heat on" pushbutton; the workpiece is heated and then either quenched in place or manually removed for quenching. A rotating spindle with holding nest can be used to hold the workpieces that need to be rotated. When a number of workpieces are heated at one time in a channel coil, non-metallic holding fixtures or trays can be used. This represents one of the lowest-cost fixturing techniques.

One position, lift fixtures can be designed for top or bottom loading. On vertical fixtures, nests are used to hold the workpieces, or the workpieces are held between centers. Different versions of horizontal fixtures exist with different types of feed mechanisms. The main difference between the one position fixture is that the workpiece is loaded outside of the coil. Depressing the "start" pushbutton causes any rotation to start and the workpiece to move into the coil. At the completion of the heat cycle, the workpiece may either quench in place or be moved into the quench.

Scanning, Workpiece Moves. The standard vertical scanner is basically a lift-and-rotate fixture that is designed to move the part progressively through the coil so that it can heat and quench progressively. In a typical cycle the workpiece is loaded into the starting position. Upon depression of the "cycle start" pushbutton, the workpiece is rapidly moved to the "heat on" position. The heat cycle starts with a timed dwell to bring the workpiece up to austenitizing temperature. The workpiece is then scanned and quenched for the desired pattern length. At the conclusion of quenching, the workpiece is moved back to the unloading/reloading position. Horizontal scanners are also used with the parts that are held and rotated on rollers. In some cases, the rollers are skewed so that the bars are moved through the coil by the rotation process. Other scanners use pushing mechanisms that push the bars through the coil. Finally, some scanners have beds or mechanisms that move the workpiece through or under the coil.

Scanning, Coil Moves. Scanners have been developed on which the coil and heat station are mounted on a mechanism that moves. The workpiece is held stationary, while the coil moves. This type of scanner is used more commonly when very heavy parts are to be induction hardened.

Higher degrees of automation use the same concepts at the coil position using index fixtures, wheeled fixtures, conveyors, shuttles, and walking-beam mechanisms to move the parts in and out of the coil. Some installations use multiple heat stations operating in sequence on the same power supply through the use of power transfer switches. Other installations heat more than one part with the same power supply, such as is done with multiple-spindle scanners, or with the use of channel- or hairpin-type coils where the workpieces can be conveyed through the coil. Considerable ingenuity exists in fixture designs with all sorts of shuttles, robotic arms, and special mechanization used.

Examples

Information was provided earlier regarding frequency selection. The Appendix "Scan Hardening" in this book gives tables for determining heating rates. However, there are some rules of thumb that can be readily used to provide a good estimate of the power selections. Note that the actual power requirements depend on the total system losses and the ability to load match the power supply to produce the required power. If in doubt, double the power required for power supply output. It is much easier to use a power supply at 50 to 75% output than to obtain 100% output power in a given coil. For the purposes of the following examples, it will be assumed that the optimum power density is being used. Table 11.6 gives some rules of thumb that will be used in the following examples for heating rates. Common rated output power supplies will then be applied. A couple of models of applications will be presented, followed by qualification of the examples and an analysis of simple hardening systems, then moving into more complex systems to increase production.

Example One: Pin Hardening. A 1045 steel pin that is 50 mm (2.0 in.) in diameter by 150 mm (6 in.) long, is to have a minimum pattern length of 100 mm (4 in.) in the center of the length induction hardened to surface hardness of HRC 50 to 60 with an effective case depth of 1 to 2 mm (0.040 to 0.080 in.). The first steps in qualification, as shown in Table 11.7, the material qualifies for both hardness and case depth. The specifications can be met with power supply frequency ranging from 10 to 450 kHz. The workpiece can be heated by single shot or scanning. Single shot, because of the power density required, will require a much higher power supply rating than scanning. Table 11.8 shows production estimates of both 10 kHz and 450 kHz, single shot heating versus scanning.

As shown, induction power supply frequencies from 10 kHz to 450 kHz will produce the required production rates. The 450 kHz will produce less total heat in the workpiece because of the shallower reference depth and lower power density required. This in turn allows the workpiece to quench

Table 11.6 Rules of thumb for estimating induction heating rates

Frequency	Case depth		Power density		Single-shot heat time, s	Scanning rate, progressive	
	mm	in.	kW/cm^2	kW/in.^{2}		mm/s	in./s
3 kHz	3.80	0.150	1.55	10	4.0	0.60	0.25
			3.10	20	2.0	1.25	0.50
10/25 kHz	2.50	0.100	1.55	10	2.5	1.00	0.40
		0.100	3.10	20	1.0	2.50	1.00
450 kHz	1.25	0.050	3.10	5	4.0	0.50	0.20
			1.55	10	2.0	1.25	0.50

The scanning rate is roughly inversely proportional to the case depth. For instance, a 10% increase in case depth will require a 10% reduction of scanning rate. Source: Ref 1

Table 11.7 Example 1

Step	Item	Description	Ref no	Analysis	Comments
1	Material	1045 cold drawn	...	...	No special or unusual geometry
2	Surface	50-mm (2-in.) diameter by 100 mm (4 in.)	...	12.56 in.^{2}	Total surface single shot
3	Hardness	HRC 55	Fig. 11.1	OK	Will quench OK
4	Case depth	1.0/2.0 mm (0.040/0.080)	Table 11.2	OK	Will quench OK
5	Min. frequency	50 mm (2.0 in.)	Table 11.2	10 kHz	...
	Opt. frequency	50 mm (2.0 in.)	Table 11.2	450 kHz	...
6	Power density	For 10 kHz	Table 11.6	20 kW/in.^{2}	Optimum
		For 450 kHz	Table 11.6	10 kW/in.^{2}	Optimum
7	Power	10 kHz	Calculated	250 kW	Needed for single shot
		10 kHz	Calculated	62 kW	Needed for scanning
		450 kHz	Calculated	125 kW	Needed for single shot
		450 kHz	Calculated	30 kW	Needed for scanning

Source: Ref 1

faster. It should be noted that the coil development of the 450 kHz, single-shot system is technically more difficult because of potential flux-intensified degradation. Both power supplies produce about the same production rate. If new equipment is being considered, the total capital equipment cost might be the deciding factor in frequency selection.

Next, consider the effect of equipment selection for this same application, but with the production requirement increased. Higher operation efficiency from mechanization will not be enough to produce this rate. Higher power, by itself, will not significantly increase the production rate unless more than one part is heated at a time. One 200 kW, 10 kHz power supply

Table 11.8 Example 1 with different processes

Heating process	Power supply		Cycle, s	Comments
Scan	100 kW, 10 kHz	Load	3	...
		Position	2	...
		Heat	5	1 s dwell plus 4 s scan
		Quench	12	...
		Return	6	at 50 mm/s (2 in./s)
		Unload	3	...
		Total	31	93/h at 80% efficiency
Single shot	250 kW, 10 kHz	Load	3	...
		Position	0	Load in position
		Heat	1	Might need slightly longer
		Quench	12	...
		Return	0	...
		Unload	3	...
		Total	19	150/h at 80% efficiency
Scan	40 kW, 450 kHz	Load	3	...
		Position	2	...
		Heat	9	1 s dwell plus 8 s scan
		Quench	10	Less heat sink, quenches faster
		Return	6	At 2 in./s
		Unload	3	...
		Total	33	87/h at 80% efficiency
Single shot	125 kW, 450 kHz	Load	3	...
		Position	0	...
		Heat	2	...
		Quench	10	Faster when quenching
		Return	0	...
		Unload	3	...
		Total	18	160/h at 80% efficiency

Source: Ref 1

can be used to scan two parts at a time. The power can be reduced from the single-shot heating as shown in Table 11.8 because the encircling coils used for scanning are more efficient than single-shot coils. The same output power, 40 kW on the 450 kHz power supply is used to operate alternately through use of a power transfer switch to two heating fixtures. The analysis of this is shown in Table 11.9. It should be noted that the operating efficiency is still estimated at 80%. It takes slightly longer to load and unload two parts at a time when done manually on the dual-spindle scanner of the 10 kHz power supply. It is very important that the materials-handling aspect of production be reviewed so that this efficiency can be maintained as the required production rate increases. The production rate of the single-shot, 250 kW, 10 kHz power supply was increased from 150 parts per hour to 164, while the rate of the 40 kW, 450 kHz power supply as single shot was increased from 87 parts per hour to 151 parts per hour. When using two heat stations with transfer switch, the power supply is kept in operation almost all of the time. As long as the heating cycle is longer than the load and unload time, the operating efficiency can be increased. The factor that controls the production rate is that a workpiece cannot be unloaded from its position until the quenching is completed. In order to optimize production rate the second heat station must be ready to have its workpiece in position for heating when the heat cycle of the first station is complete. With RF, about 1 s is needed for the actual power transfer.

Example Two: Gear Hardening. A fine-pitch steel gear made from 1144 steel that is 5.0 mm (2.0 in.) in diameter by 12.5 mm (0.50 in.) wide is to be induction hardened. The gear teeth are to be hardened to a surface hardness of HRC 50/55 and a total case depth of 1 mm (0.040 in.) minimum below the root. The qualification is shown in Table 11.10. The power supply frequency can range from 25 kHz to 450 kHz. A 50 kW, 25 kHz and a 30 kW 450 kHz power supply have been selected.

The analysis will first start with the simplest fixturing or handling equipment. Then examples will show how higher production rates can be

Table 11.9 Example 1 for higher productivity

Induction process	Power supply		Seconds	Comments
Scan (2 parts)	200 kW, 10 kHz with dual-spindle scanner	Load	6	...
		Position	2	...
		Heat	5	1 s dwell plus 6 s scan
		Quench	12	...
		Return	6	At 2 in./s
		Unload	4	...
		Total	35	164/h at 80% efficiency
Scan	40 kW, 450 kHz two station with transfer switch for alternating cycle	Position	0	8 s needed to unload and reload; part loaded during cycle of opposing station; 1 s dwell plus 8 s scan
		Heat	9	
		Quench	10	
		Return	6	
		Transfer	1	
		Total	19 average	151/h at 80% efficiency

Source: Ref 1

Table 11.10 Example 2

Step	Item	Specification	Ref no.	Analysis	Comments
1	Material	1144 cold drawn	...	...	...
2	Surface	2 in. diam by 0.50 in. wide	...	1.57 in.^{2}	Surface to be heated
3	Hardness	HRC 50	Fig. 11.1	OK	Will need tempering
4	Case depth	0.040 in. (1.0 mm) min	Table 11.2	OK	...
5	Min freq.	2.0 in. (50 mm) diam	Table 11.2	25 kHz	...
	Opt freq.		Table 11.2	450 kHz	...
6	Power density	For 25 kHz	Table 11.6	15 kW/in.^{2}	Optimum
		For 450 kHz	Table 11.6	10 kW/in.^{2}	Optimum
7	Power	25 kHz	Calculated	50 kW	Rec. power supply
		450 kHz	Calculated	30 kW	Rec. power supply

Rec., recommended. Source: Ref 1

produced by changing the fixturing and power supply ratings. Table 11.11 shows the production rate that can be produced with simple fixturing consisting of a single shot, lift and rotate operating with 25 kHz, 50 kW and 450 kW, 30 kW power supplies, each with lift-and-rotate fixtures. At this point, production rate of both frequencies is about the same. A three-second, static heat cycle is about correct for this steel. The 25 kHz power supply would not be operating at full power to produce a 3 s heat cycle, but as mentioned the power supply should not be rated too close to the actual power requirements. Total system losses for both frequency power supplies would vary a little but not significantly enough to be a factor in equipment selection. Other factors for consideration might be the reduced maintenance requirements for solid state, plus the type of other applications that might be run on the equipment. Next, examples will be given for increasing the productivity of the same power supplies on the same workpiece through mechanization and different set up. The effect of running two workpieces at a time on the 450 kHz power supply versus operating the 450 kHz with two alternating heat stations is presented in Table 11.12. In either operation, one operator is used. The production rate is slightly higher on the 450 kHz unit, but probably not high enough to justify the two-position fixturing. The two-position becomes more effective when the

Table 11.11 Example 2 with simple fixturing

Induction process	Power supply		Seconds	Comments
Single shot	50 kW, 25 kHz	Load	3	...
lift and rotate		Position	1	...
		Heat	3	...
		Quench	6	...
		Drop	1	...
		Unload	2	...
		Total	16	180/h at 80% efficiency
Single shot	30 kW, 450 kHz	Load	3	
lift and rotate		Position	1	
		Heat	3	
		Quench	6	
		Drop	1	
		Unload	2	
		Total	16	180/h at 80% efficiency

Source: Ref 1

heating and quenching times are long enough that the power supply can be effectively kept in production more than 90% of the time. Index tables make up another type of fixturing that can be used to further increase the production.

Index fixtures are a fairly simple way to increase the production rate running one or two parts at a time. An index fixture requires that the workpiece be either manually loaded or mechanically loaded, indexed into the coil location, heated, quenched, and moved out of the coil for unloading (either manual or mechanical). The overall cycle can be replicated on a timed basis, which tends to improve the operating efficiency because the operators have to maintain the pace set by the process. Processes that are totally mechanized can approach 90 to 95% operating efficiency due to the ability to keep the induction power supply in production consistently.

The design recommendations for fixtures are discussed in the Chapter "Induction Heat Treating Systems" in this book. Index fixtures vary from rotary index tables to conveyorized and walking beam mechanisms. Index tables are more common, and rotation is usually needed at the coil and primary quenching positions. The fixtures can be designed for a second quenching position. Quenching can be done either in or below the coil until the workpiece is quenched to a temperature sufficiently low that the quenching can be completed when the workpiece is then indexed to the second position. The second quenching may be done primarily so that either the operator can handle the workpiece or the workpiece is cooled sufficiently for in-line induction tempering next. The use of a secondary quench also permits the workpiece to be removed from the coil position so that another workpiece can be indexed into the coil and the heat cycle initiated while final quenching is done on the previous part at the secondary quench position. Using the same gear application and requirements, changing the fixturing to a rotary index

Table 11.12 Example 2 with more complex fixturing

Induction process	Power supply		Seconds	Comments
Single shot	50 kW, 25 kHz	Load	4	...
(2 parts)	one spindle	Position	1	...
		Heat	3	...
		Quench	6	...
		Drop	1	...
		Unload	4	...
		Total	19	303/h at 80% efficiency
Single shot	30 kW, 450 kHz	Heat	3	Loading and unloading done
	two station with transfer	Quench	6	during heating of other station
	switch for alternating cycle	**Total:**	9	
		Unload	2	
		Load	3	
		Position	1	
		Drop	1	
		Total	7	320/h at 80% efficiency

Source: Ref 1

fixture will increase the production on the same power supply to the required level of 480 workpieces per hour. Table 11.13 shows the production increase from an index fixture.

Use of an index fixture increases the production rate higher than that obtained by a lift-and-rotate fixture. Also, as mentioned, index fixtures, because of their mechanization and ability to run faster through the use of secondary quenches, can operate at a high efficiency. An increase of 80% to 90% efficiency is expected with optimization of workpiece production flow.

Example 2 involved static heating. Scanners are used when progressive heating is required. Scanners are built for both vertical and horizontal hardening. Vertical scanners hold the workpieces either in a bottom nest or between vertical spindles, and heat and quench the workpiece on a continuous basis as it is passed through the induction coil. As previously mentioned, through the use of the scanning technique, the power requirements of the power supply are minimized. This technique is necessary for very long workpieces because the power requirements are not feasible. Rotation of the parts is generally required in a scanner. When the workpiece length is excessive, scanners may be designed for horizontal scanning rather than vertical scanning. There are other applications in which horizontal scanners are used because the hardened workpieces tend to be straighter, particularly on small diameter bars.

Scanning Application. Instead of scanning a straight shaft, the application of the gear hardening will be used to show the increase in production that can be obtained through fixture change. When scanning gears on rods, good setup techniques need to be used to make certain that uniform cases and hardnesses are produced at the start and the end of the heating cycle. The gears must have precise dwells when the heating cycle is initiated, and the coil must be designed so that when the heat is turned off at the end of the cycle, the trailing edge does not overheat while leaving a soft zone just before the edge. Furthermore, someone other than the operator must load the gears on the rods because the operator cannot load them fast enough. Table 11.14 shows an example of the high production rate that can be produced.

The previous examples are meant to show the versatility of processing that can be accomplished. With good definition of the production require-

Table 11.13 Example 2 with index fixture

Induction process	Power supply		Seconds	Comments
Static heating, index, 2 gears	30 kW, 450 kHz	Load	0	Loading is done during heating and quenching
		Position	3	
		Heat	4	
		Drop	1	
		Quench	4	Plus 4 s secondary quench
		Total	**12**	**480/h at 80% efficiency**

Source: Ref 1

ments, there may be many ways that the production of the same workpiece can be accomplished.

At this point it should be noted that from the same power supply, through more complex fixturing, the production rate has been increased from 180 parts per hour to 886 parts per hour. Sufficient organization and planning have been done so that workpieces can be taken from storage or the prior manufacturing process and removed from the area. For instance, as mentioned, one operator cannot load the gears on rods fast enough to keep the scanner in operation. Furthermore, the types of skids or pallets used to store the workpiece and even the workpiece stacking and orientations can affect production.

Vertical scanners may have one or more spindles holding workpieces, so that more than one workpiece can be hardened at a time. While there is no theoretical number of heating positions, the loading and unloading of the workpieces dictates the economical number to be run at a time. Generally a dual-spindle scanner becomes more economical if the operator can load and unload two workpieces at a time. Because heavier parts are most often loaded one at a time, it may be more economical to run these on only one spindle.

There are many other types of mechanical handling devices of which the only limit is ingenuity. Anything that has the ability to move parts into the coil without being affected itself by the magnetic flux field of the coil can be used. Examples or other types are trap fixtures, bowl feeders, conveyors, rotating wheels, shuttles, and cam operated devices.

Tempering will be required if the parts, as with the prior gear illustrations, quench to a hardness higher than the specification. The theory of tempering is discussed in the Chapter "Tempering" in this book. Workpieces may be tempered by induction or furnace. If tempered by induction, tempering on the same power supply will slow production substantially because tempering heat cycles are generally at two to three times the cycle time required for austenitizing. (As previously discussed, slower heat input is needed to provide surface to core uniformity of heat.) If tempering can be done with the same power supply, coil, and fixture, and if reduction of the overall production rate is not a considera-

Table 11.14 Example 2 with scanner

Induction process	Power supply		Seconds	Comments
Scanning with gears on rods (16 gears)	50 kW, 25 kHz	Load	5	...
		Position	3	...
		Heat	9	1 s dwell plus scan 8 in. at 1 in./s
		Quench	12	...
		Return	5	...
		Unload	4	...
		Total	39	(a) 886/h at 60% efficiency

(a) The operating efficiency will tend to be lower because of the time required to handle the gears on rods. Source: Ref 1

tion, then the overall cycle adjustment for tempering can be made. If not, a second power supply and coil can be used, sometimes transferring the part so that the tempering is done as the workpiece is sequenced through a series of coils. For instance, if a tempering time of 24 s is needed, tempering the workpiece at 8 s each in 3-coil positions will produce the overall 24 s. In fact, the tempering might be more uniform because the transfer time between the coils produces some soaking and temperature equalization.

Post production analysis must consider the removal of quenchant from the workpieces before tempering. Oils must be removed. Polymers, as previously discussed in the Chapter "Heat Treating Basics" in this book, when heated produce a strong, noxious smell. In addition, if the burning is not complete, the polymers bake on in hard-to-remove deposits. Various washing processes are available. Standard hot water washers will remove quenchants. Disposal of the removed quenchants from the washer is an issue that must be addressed in accordance with environmental regulations.

Rust preventive application and packaging of the parts that were hardened is a final consideration. The parts, as removed from tempering, are very susceptible to rust. Decisions should be made as to how the rust preventive will be applied, such a spraying or dipping.

Setup Instructions and Procedures. Replication of the processing is important, whether a power supply and heating system are in production continuously on one application or whether many set-ups are made over a period of time. Written process instructions must be made including all specifications, prints, coils, fixtures, tuning instructions, specific process instructions, and inspection procedures, and requirements. An example of the instructions for set up personnel follows (if the operator is doing the set up, then these would apply to the operator).

Setup Instructions. Good setup is critical for successful induction heat treating. The setup person is responsible not only for producing a first-article part, which is completely in specification, but also for making a setup on which an operator can safely run the parts on a production basis. The setup person is responsible for making setups which provide optimum production rates. Finally, the setup person is responsible for operator training and for timely response to any production problem.

Setup Procedures are as follows.
1. Read work order, set up sheet, prints, and any other instructions enclosed in the work order packet. Make certain everything is clear and all information needed is on hand. Verify the core hardness is correct if there is a core hardness requirement, and record it on work order.

2. Remove and take the old setup tooling to tool room for cleanup and inspection. Find the tooling on the new set up as indicated on the setup sheet. Bring the tooling and parts to the machine.
3. Check the quenchant concentration and condition so that if a change is needed, it can be done while the setup is in process. Record quenchant concentration on setup sheet.
4. If the job has been processed before, make the setup according to instructions. If the setup has not been run before, find a similar part for initial tuning of the power supply. If similar tuning cannot be found, tune according to the instructions of the power supply being used, starting out at low power.
5. Make certain the coil attachment point on the buss or transformer is clean before installing the induction coil. Turn on the coil water. Check the coil for water flow and for leaks.
6. Install adapters and fixtures, as specified. Make certain coil is centered on workpiece.
7. If information is needed regarding operating instructions for a given machine or fixture, check the operating manual at the machine or in the engineering office.
8. Check the setup sheet for control settings. If the proper program for that machine is indicated, as with numerical control (NC) fixtures, enter the program. If a new NC program needs to be entered, begin in the following order:
   - Establish start position, heat position, quench position, and index position if applicable (most often the same as the start position).
   - Enter the scan speed if it is a scanning application (note that slightly too fast is best if in doubt), set a power level (start low and increase as needed). Enter an approximate heat time, remove the workpiece, and run a cycle (no heat) to make certain it runs properly.
   - Put a setup piece in place and try a cycle with heat and quench.
   - Adjust the scanning speed, power, and/or time for optimum heating.
9. Run first article part.
10. Check the part for conformance on hardness, length, case depth, surface finish, etc., and qualify the first article.
11. Record all setup information if new. Indicate any changes, if made, to a previous setup.
12. After the first article part is qualified as in specification, let production know the setup is ready for production.

Stop production immediately on any induction heater that is in production and that has production potentially in nonconformance. If a nonconformance exists, any nonconforming parts must be identified and segregated. Work with production control and quality control to investigate and remedy any problems with non-conforming parts. Make suggestions for continuous improvement as appropriate. Report any injuries to the shift foreman promptly.

Operator Instructions. The induction heater operator is very important to the induction heat treating process. The operator is a critical link in both production control and quality control. Operators have the following responsibilities:

- Make certain the quality control has approved the part for production.
- Make certain that all operating instructions are clearly understood. (Questions may be asked any time before, during, or after processing to help in understanding induction heat treating.)
- Handle parts carefully. Parts with machined surfaces such as gears, threads, and bearings are not to be dropped on another part.
- Make certain that all parts loaded onto spindles or into fixtures are properly located.
- Shut off the induction heater if any type of operating problem or induction heater fault is suspected. This includes parts not properly locating, coil arcs, or any change of process. This also includes detecting any problem that might be in the parts before hardening, such as cracks or poor machining.
- Make certain that parts which are partially heat treated (because of a machine fault), and/or parts that have not been heat treated are not mixed with parts that have been heat treated.
- Learn and memorize the heat color (temperature) on the part being heated. Look at each part after quenching, whenever possible, to make certain the part has been heat treated, that the discoloration is as expected in both the heated and unheated areas, and that there are no visible cracks or melts.
- Follow all safety regulations and practices. Advise anyone in management about any condition you feel is unsafe. Use proper and safe work-handling techniques when handling parts and skids of parts.
- Complete and attach job/flow tags to containers of hardened parts before they are transported.
- When the order is complete, verify the order is complete. If there is a count difference, note it on the work order. Verify that there are no unprocessed parts inside the containers in which the parts were sent to induction hardening.

REFERENCES

1. R.E. Haimbaugh, Induction Heat Treating Corp., personal research
2. Standard Practices and Procedures for Steel, *Heat Treaters Guide*, American Society for Metals, 1982

3. *Heat Treating*, Vol 4, *ASM Handbook*, ASM International, 1991
4. C.A. Tudbury, *Basics of Induction Heating*, John F. Rider, Inc., New Rochelle, NY, 1960
5. V. Rudnev et al., *Steel Heat Treatment Handbook*, Marcel Dekker, Inc., 1997

CHAPTER **12**

Standards and Inspection

INSPECTION IS required in order to provide assurance of conformance to process specifications. This chapter will cover both the relevant standards and practices, and inspection techniques for induction-hardened parts. Inspection involves two different types of testing, destructive (DT) and nondestructive (NDT). The names define what they mean: for instance, when a workpiece is destructively tested, the workpiece is destroyed for practical use. Specifications for induction-hardened parts will always require some sort of hardness measurement and may require case-depth verification, microstructural requirements, and possibly some sort of physical measurement such as straightness.

Hardness testing may be destructive or nondestructive, depending upon the hardness scale used and the method of preparation for testing. For example, a Rockwell C indent on a final ground surface may be detrimental to further use of the part, whereas a Rockwell 15N indent at the same location may not roughen the surface finish enough to prevent use. Any preparation such as grinding the surface may be destructive, and of course a part is destroyed when sectioned. While most hardness testing is done by use of a hardness tester that produces an indent on the part, there are still a few specifications that incorporate hardness testing through use of test files. Table 12.1 compares the advantages and limitations of nondestructive and destructive testing.

Standards

Reference 2 provides standards and practices, as well as recommendations for inspection. Workpiece hardness is always verified. The ASTM standard for hardness testing is the current version for Rockwell testing, ASTM E18-88 and for microhardness testing, ASTM 384. Rockwell testers are most commonly used, while microhardness testers are used for specific applications. The Chapter "Heat Treating Basics" in this book

Table 12.1   Advantages and limitations of Nondestructive Testing (NDT) and Destructive Testing (DT)

Nondestructive testing	Destructive testing
Advantages	**Advantages**
• Can be done directly on production items without regard to part cost or quantity available, and no scrap losses are incurred except for bad parts • Can be done on 100% of production or on representative samples • Can be used when variability is wide and unpredictable • Different tests can be applied to the same item simultaneously or sequentially • The same test can be repeated on the same item. • May be performed on parts in service • Cumulative effect of service usage can be measured directly. • May reveal failure mechanism • Little or no specimen preparation is required. • Equipment is often portable for use in field. • Labor costs are usually low, especially for repetitive testing of similar parts.	• Can often directly and reliably measure response to service conditions • Measurements are quantitative, and usually valuable for design or standardization. • Interpretation of results by a skilled technician is usually not required. • Correlation between tests and service is usually direct, leaving little margin for disagreement among observers as to meaning and significance of test results.
Limitations	**Limitations**
• Results often must be interpreted by a skilled, experienced technician. • In absence of proven correlation, different observers may disagree on meaning and significance of test results. • Properties are measured indirectly, and often only qualitative or comparative measurements can be made. • Some nondestructive tests require large capital investments.	• Can be applied only to a sample, and separate proof that the sample represents the population is required • Tested parts cannot be placed in service. • Repeated tests of same item are often impossible, and different types of tests may require different samples. • Extensive testing usually cannot be justified because of large scrap losses. • May be prohibited on parts with high material or fabrication costs, or on parts of limited availability • Cumulative effect of service usage cannot be measured directly, but only inferred from tests on parts used for different lengths of time. • Difficult to apply to parts in service, and usually terminates their useful life • Extensive machining or other preparation of test specimens is often required. • Capital investment and manpower costs are often high.

Source: Ref 1

discusses the general interpretations of hardness. Case-depth measurement is discussed later in the Chapter. The ASTM *Annual Book of the American Society of Testing and Materials* (Ref 3) publishes an annual volume on NDT called *Volumes 03.03 Nondestructive Testing* that covers many NDT procedures.

Test Equipment

Hardness Testing. The hardness scale specified on the print should be used unless inappropriate. The most common condition found where a scale might be inappropriate is one in which the part being tested or the case depth is not thick enough to support the test scale being used. As an example, most prints specify hardness using the Rockwell "C" scale. However, the "C" test scale is inappropriate for very small rounds because an accurate indent cannot be made. Hardness scales are also inappropriate in cases where the hardness being tested is less than ten times the depth of the indenter penetration. Table 12.2 should be used to determine the minimum case or cross-sectional thickness needed to support different hardness scales at various hardness test

Table 12.2 Minimum cross-section size to support hardness scale

Thickness		Rockwell scale		
		A		C
mm	Inch	Dial reading	Approximate hardness C-scale	Dial reading
0.36	0.014	...	...	...
0.41	0.016	86	69	...
0.46	0.018	84	65	...
0.51	0.020	82	61.5	...
0.56	0.022	79	56	69
0.61	0.024	76	50	67
0.66	0.026	71	41	65
0.71	0.028	67	32	62
0.76	0.030	60	19	57
0.81	0.032	...	...	52
0.86	0.034	...	...	45
0.91	0.036	...	...	37
0.96	0.038	...	...	28
1.02	0.040	...	...	20

Thickness		Rockwell Superficial Scale					
		15N		30N		45N	
mm	Inch	Dial reading	Approximate hardness C-scale	Dial reading	Approximate hardness C-scale	Dial reading	Approximate hardness C-scale
0.15	0.006	92	65	...	...	...	...
0.20	0.008	90	60	...	...	...	...
0.25	0.010	88	55	...	...	...	...
0.30	0.012	83	45	82	65	77	69.5
0.36	0.014	76	32	78.5	61	74	67
0.41	0.016	68	18	74	56	72	65
0.46	0.018	...	...	66	47	68	61
0.51	0.020	...	...	57	37	63	57
0.56	0.022	...	...	47	26	58	52.5
0.61	0.024	...	...	...	...	51	47
0.66	0.026	...	...	...	...	37	35
0.71	0.028	...	...	...	...	20	20.5
0.76	0.030	...	...	...	...	...	...

A guide for selection of scales using the diamond penetrator. Source: Ref 2

ranges. The "C" scale has the heaviest weight, penetrates the deepest, and tends to give the most accurate readings for a variety of surface conditions and parts. Some of the general ASTM standards important to testing are:

- *Distance between Indents:* Hardness indents must be three indentation diameters apart.
- *Distance to the Edge of a Part:* Hardness indents must be 2.5 indentation diameters from the edge of a part.
- *Number of Readings for a Good Test:* The recommended number is three valid readings. If a part rocks or has movement, the reading should be discarded and another reading should be taken.
- *Round Correction:* When a round or curved workpiece is tested, there are two sources of error. First, the Rockwell tester being used

has what is called *anvil error*. The diamond indenter may flex just a little bit when a round is measured, moving it off the true centerline. This type of error can vary from testing machine to testing machine. If anvil error is in question, measure the hardness on two different hardness scales. The lighter test scale will tend to have less error, and comparison of the results of the two different hardness scales can indicate whether there is testing error. Second, because a round is being tested instead of a flat, the depth of penetration is slightly different. Above 25 mm (1 in.) in diameter, the tests are fairly accurate. However, below this round, correction must be added to the test readings. Table 12.3 shows the table for determining the round correction that should be added to the reading. Note that parts should not be tested below 6.25 mm (¼ in.) in diameter on the "C" and "A" scale and below 3.12 mm (⅛ in.) in diameter on the 15N test scale. Flats should be ground on small diameters to provide accurate testing surfaces.

Always record any test records when round corrections are used. If the test scale is being converted to another scale, always add the round correction to the test result before correlating to another scale. Finally, always note on inspection records the scale from which the conversion was made. Correlation is not always totally accurate and is intended to give an approximation of readings from one test scale to another. Figure 12.1 shows the use of steady rests for long workpieces. Figure 12.2

Table 12.3 Round correction tables

C D A Scale	Diamond "brale" penetrator Diameter of specimen, inches						
	¼	⅜	½	⅝	¾	⅞	1
80	0.5	0.5	0.5	0	0	0	0
70	1.0	1.0	0.5	0.5	0.5	0	0
60	1.5	1.0	1.0	0.5	0.5	0.5	0.5
50	2.5	2.0	1.5	1.0	1.0	0.5	0.5
40	3.5	2.5	2.0	1.5	1.0	1.0	1.0
30	5.0	3.5	2.5	2.0	1.5	1.5	1.0
20	6.0	4.5	3.5	2.5	2.0	1.5	1.5
15 N 30 N 45 N Scale	Diamond "N" brale penetrator Diameter of specimen, in.						
	¼	⅜	½	⅝	¾	⅞	1
90	0.5	0.5	0	0	0	0	0
85	0.5	0.5	0.5	0	0	0	0
80	1.0	0.5	0.5	0.5	0	0	0
75	1.5	1.0	0.5	0.5	0.5	0.5	0
70	2.0	1.0	1.0	0.5	0.5	0.5	0.5
65	2.5	1.5	1.0	0.5	0.5	0.5	0.5
60	3.0	1.5	1.0	1.0	0.5	0.5	0.5
55	3.5	2.0	1.5	1.0	1.0	0.5	0.5
50	3.5	2.0	1.5	1.0	1.0	1.0	0.5
45	4.0	2.5	2.0	1.0	1.0	1.0	1.0
40	4.5	3.0	2.0	1.5	1.0	1.0	1.0

Source: Ref 2

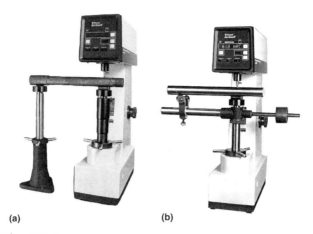

Fig. 12.1 Steady rest support for long workpieces. Source: Ref. 4

shows the correct use of a v-block anvil for checking a round part. Workpieces that have irregular test surfaces may need to be cross-sectioned so that the surface hardness can be tested just beneath the surface on the cut cross section.

Microhardness testing, as discussed in the Chapter "Heat Treating Basics" in this book, is not used as widely for induction-hardened parts as Rockwell testing. Most specifications for case depths do not require microhardness. When microhardness is required, the Vickers indent is recommended. This makes testing of the HRC 50 equivalent easier in the relatively large hardness transition zones produced during induction hardening. With the Vickers, polishing to 600 grit will produce a specimen

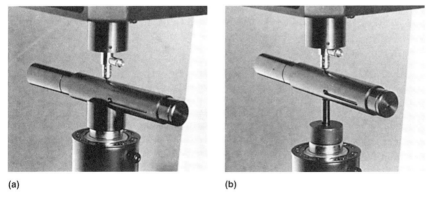

Fig. 12.2 Anvil support for cylindrical workpieces. (a) Correct method places the specimen centrally under indenter and prevents movement of the specimen under testing loads. (b) Incorrect method of supporting cylindrical work on spot anvil. The testpiece is not firmly secured, and rolling of the specimen can cause damage to the indenter, or erroneous readings. Source: Ref 4

surface smooth enough for a good test. (The actual standards for metallographic test specimen surface preparation are given in ASTM E3.) Some powdered metal hardness specifications permit the use of apparent hardness tests that read the average hardness as tested rather than the true hardness of the matrix. Average hardness is then correlated, through tests, to the core hardness and listed in their specifications. For instance, a test producing 35 Rockwell C may indicate a matrix hardness when tested by microhardness of 50 Rockwell C as correlated from Vickers to Rockwell C. Powdered metals, because of porosity, may require clusters of indentations made with obvious low readings discarded and the other readings averaged. Microhardness testing is necessary for accurate testing of the cast irons that have graphite, such as ductile cast iron, and microhardness is also used for the measure of case depths on automotive parts such as wheel hubs.

Microstructures can be seen through a microscope after workpieces are properly prepared and etched. The preparation techniques will be discussed later in this Chapter.

Nondestructive Testing (NDT)

Nondestructive testing is used for quality-control purposes for the obvious intention of not destroying good parts while testing and evaluating for conformance to specifications. Many quality considerations must be applied to the control on NDT processes to ensure that the information being supplied from them is accurate, timely, and germane. One of the greatest problems of nondestructive inspection has been misapplication, which usually means that the wrong information was supplied. Louis Cartz, in the ASM book *Nondestructive Testing* (Ref 5), cites that liquid penetrant and magnetic particle inspection account for about 50% of all NDT testing, with ultrasonic and x-ray methods accounting for about another third. Eddy current accounts for approximately 10%. All other methods were about 2%. Table 12.4 provides examples on the relative uses and merits of various NDT methods from an operational viewpoint, while Table 12.5 provides a comparison of the same NDT methods for the characteristics tested. Eddy current, magnetic particle, and liquid penetrant are the most widely used methods for testing induction-hardened parts.

Successful application of NDT methods to the inspection of induction-hardened parts requires that:

- The test system and procedures are suited to both the inspection objectives and types of flaws to be detected
- The operator has sufficient training and experience
- The standard for acceptance appropriately defines undesirable characteristics of a nonconforming part.

Table 12.4 Relative use and merits of various nondestructive testing methods

	Test method				
	Ultrasonics	x-ray	Eddy-current	Magnetic particle	Liquid penetrant
Capital cost	Medium to high	High	Low to medium	Medium	Low
Consumable cost	Very low	High	Low	Medium	Medium
Time of results	Immediate	Delayed	Immediate	Short delay	Short delay
Effect of geometry	Important	Important	Important	Not too important	Not too important
Access problems	Important	Important	Important	Important	Important
Type of defect	Internal	Most	External	External	Surface breaking
Relative sensitivity	High	Medium	High	Low	Low
Formal record	Expensive	Standard	Expensive	Unusual	Unusual
Operator skill	High	High	Medium	Low	Low
Operator training	Important	Important	Important	Important	
Training needs	High	High	Medium	Low	Low
Portability of equipment	High	Low	High to medium	High to medium	High
Dependent on material composition	Very	Quite	Very	Magnetic only	Little
Ability to automate	Good	Fair	Good	Fair	Fair
Capabilities	Thickness gauging; some composition testing	Thickness gauging	Thickness gauging; grade sorting Case depth hardness	Defects only	Defects only

Source: Ref 5

Table 12.5 Comparison of some nondestructive testing methods

Method	Characteristics detected	Advantages	Limitations	Example of use
Ultrasonics	Changes in acoustic impedance caused by cracks, nonbonds, inclusions, or interfaces	Can penetrate thick materials; excellent for crack detection; can be automated	Normally requires coupling to material either by contact to surface or immersion in a fluid such as water. Surface needs to be smooth.	Adhesive assemblies for bond integrity; laminations; hydrogen cracking
Radiography	Changes in density from voids, inclusions, material variations; placement of internal parts	Can be used to inspect wide range of materials and thicknesses; versatile; film provides record of inspection	Radiation safety requires precautions; expensive; detection of cracks can be difficult unless perpendicular to x-ray film.	Pipeline welds for penetration, inclusions, voids; internal defects in castings
Visual-optical	Surface characteristics such as finish, scratches, cracks, or color; strain in transparent materials; corrosion	Often convenient; can be automated	Can be applied only to surfaces, through surface openings, or to transparent material	Paper, wood, or metal for surface finish and uniformity
Eddy-current	Changes in electrical conductivity caused by material variations, cracks, voids, or inclusions	Readily automated; moderate cost	Limited to electrically conducting materials; limited penetration depth	Heat exchanger tubes for wall thinning and cracks
Liquid penetrant	Surface openings due to cracks, porosity, seams, or folds	Inexpensive, easy to use, readily portable, sensitive to small surface flaws	Flaw must be open to surface. Not useful on porous materials or rough surfaces	Turbine blades for surface cracks or porosity; grinding cracks
Magnetic particles	Leakage magnetic flux caused by surface or near-surface cracks, voids, inclusions, material or geometry changes	Inexpensive or moderate cost, sensitive both to surface and near-surface flaws	Limited to ferromagnetic material; surface preparation and post-inspection demagnetization may be required	Railroad wheels for cracks; large castings

Source: Ref 5

Indications obtained during NDT testing need to be interpreted and evaluated. Any indication that is found is called a discontinuity. Discontinuities are not necessarily defects, but they need to be identified and evaluated to decide whether the part is at or below specification. Table 12.6 lists the various NDT processes and the types of defects that can be indicated.

Magnetic particle inspection is probably the most versatile in the types of defects that can be revealed. The magnaflux tester is used for this type of inspection. Magnetic-particle inspection is a method for locating surface and subsurface discontinuities in magnetic steels. When a part under test is first magnetized with any magnetic discontinuities, cracks that lie in a direction that is generally transverse to the direction of the magnetic field cause a leakage field to be formed at and above the surface of the part. Next finely divided ferromagnetic particles are applied over the surface to hold some of the particles in place over the cracks. When ultraviolet light (black light) is applied to the parts, the forms will fluoresce with their outline generally indicating crack location, size, shape, and extent. Small and shallow surface cracks greater than 0.125 mm (0.005 in.) can be detected, as long as the part is located in a magnetic field that will intercept the crack. Cracks below the surface are difficult to detect.

Magnafluxing is operator dependent, and the reliability is not always 100% because interpretation is subjective. Inspection depends on the good eyesight, acuity, and experience of the inspector. Where a powder indication of uneven shape occurs, the inspector may need to repeat the process to see if the indication is reproducible. Some general comments are:

- Surface defects tend to give rise to sharp, narrow particle patterns held tightly together. The deeper the crack, the greater the buildup.
- Subsurface defects give rise to broad, fuzzy particle patterns because the magnetic particles do not adhere as tightly.

False indications are not the result of magnetic forces. Some patterns of particles are formed by surface roughness or are held together mechanically. These indications will often not reappear on reprocessing. False indications can result from the presence of lint, hair, fingerprints, or surface tension drain patterns. Resulphurized steels such as American Iron and Steel Institute (AISI) 1144 will have particle build-ups around the surface sulphide inclusions. Nonrelevant indications are caused by flux leakage that has no relation to a discontinuity that is considered to be a defect.

Accumulations of powder occur at corners, where there is a change in the cross section of the part or change of shape. Care and discretion must be exercised because cracks are also likely to occur in these regions.

Table 12.6 Selection of nondestructive testing process (visual examination is always useful)

Defect	Method	Comments
Bursts (wrought metals)	Ultrasonic testing	Internal bursts produce a sharp reflection; able to differentiate types of bursts
	Magnetic particle testing	Surface and near-surface bursts only of ferromagnetic materials
Cold shuts (casts)	Liquid penetrant inspection	Surfaces of most metals; smooth regular line; casts difficult
	X-radiography	Distinct dark line
Fillet cracks (bolts) (wrought metals)	Ultrasonic testing	Extensively used; sharp reflection
	Liquid penetrant inspection	All metals; sharp, clear indications; necessary to remove all penetrant subsequent to testing
Grinding cracks	Liquid penetrant inspection	All metals; irregular pattern fine cracks; may require long penetrant dwell times
	Magnetic particle inspection	Ferromagnetic metals only
Convolution cracks nonferrous	X-radiography	Extensively used
Heat-affected zone cracking (HAZ)	Magnetic particle inspection	Ferromagnetic metals only; demagnetization may be difficult; must avoid electric arc from prods
	Liquid penetrant inspection	Nonferrous welds; depends on surface processing
Heat-treat cracks (near areas of stress)	Magnetic particle inspection	Ferromagnetic materials; straight, forked, or curved indications
	Liquid penetrant inspection	Nonferrous metals
Surface shrink cracks	Liquid penetrant inspection	Nonferrous metals; avoid regions such as press fittings
	Magnetic particle inspection	Ferromagnetic materials
	Eddy-current	Nonferrous welded piping
Thread cracks (wrought metals)	Liquid penetrant inspection	Use fluorescent penetrant
	Magnetic particle inspection	Ferromagnetic metals; nonrelevant indications from threads
Tubing cracks (nonferrous)	Eddy-current	Recommended if tube diameter is less than 1 in. and wall thickness is less than 0.15 in.
	Ultrasonic testing	Suitable for tubing; couplants may affect certain alloys
Hydrogen flake (ferrous)	Ultrasonic testing	Extensively used
	Magnetic particle inspection	Used on finished part; appearance of short discontinuities (hairline cracks)
Hydrogen embrittlement (ferrous)	Magnetic particle inspection	Indications appear as randomly oriented cracks
Inclusions (welds)	X-radiography	Used extensively; sharp well-defined round or other shaped spots; relatively large inclusions
	Eddy-current	Thin wall welded tubing
Inclusions (wrought metals)	Ultrasonic testing	Used extensively; large inclusions act as good reflectors; smaller give rise to background "noise"
	Eddy-current	Thin-wall, small diameter rods; difficult for ferromagnetic metals
	Magnetic particle inspection	On machined surfaces, indicators are straight intermittent or continuous line
Lack of penetration (welds)	X-radiography	Extensively used
	Ultrasonic testing	Used but some geometries difficult
	Eddy-current	Nonferrous welded tubing
	Magnetic particle inspection; liquid penetrant inspection	If rear of weld is visible
Laminations (wrought metals)	Ultrasonic testing	Extensively used; sharp signals with loss of rear wall signal
	Magnetic particle inspection; liquid penetrant inspection	Indication straight, broken lines
Laps and seams (rolled metals)	Liquid penetrant inspection; magnetic particle inspection (ferrous)	All metals; fluorescent LPI; indications curved, continuous or broken lines
Laps and seams (wrought metals)	Magnetic particle inspection (ferrous); liquid penetrant inspection	Straight, spiral, or curved indications
	Ultrasonic testing	Extensively used; good signals
	Eddy-current	Tubing and piping
Microshrinkage (magnesium castings)	X-radiography	Extensively used; elongated feathery streaks
	Liquid penetrant inspection	Extensively used on finished surfaces where machining opens micropores
Gas porosity (welds)	X-radiography	Extensively used; round and elongated spots on radiograph
	Ultrasonic testing	Very sensitive, but depends on grain size
	Eddy-current	Thin wall tubing
Unfused porosity (aluminum)	Ultrasonic testing	Extensively used
	Liquid penetrant inspection	Machined article; straight line indications
Stress-corrosion cracking	Liquid penetrant inspection	Extensively used
Hot tears (ferrous coatings)	X-radiography	Extensively used
	Magnetic particle testing	Surface only
Intergranular corrosion (nonferrous)	Liquid penetrant inspection	Extensively used
	X-radiography	Advanced stages of intergranular corrosion

Source: Ref 5

Magnetic labeling and external magnetic fields occur when a part is partially magnetized by a neighboring electric current or by a permanent magnet. Sharp indications may result, though these will disappear after demagnetization. Residual magnetization may result from prior use of a magnetic chuck, and this will usually give a fuzzy indication. In this case, demagnetization is required first. Sharp corners, thread roots, keyways, and fillets give rise to a magnetic leakage field that can produce fuzzy indications. The magnetizing force should be reduced, and the process should be repeated. Metal path constrictions occur when size reductions give rise to apparent magnetic leakage fields with resulting fuzzy indications. Deep machine grooves can accumulate powder residue during draining; this provides false indications. One method of verification is to gently wipe off the excess and see if the indication reappears.

After establishing a suitable magnafluxing technique for inspection of a particular part, the process must then be maintained by adequate controls. These controls include periodic calibration of equipment, maintaining an adequate concentration of magnetic particles in the carrier solution, uniformity of pre-inspection part preparation, careful attention of post-inspection requirements such as removal of residual magnetism, and, most important of all, periodic evaluation of personnel to maintain a suitable level of proficiency.

Liquid-penetrant inspection is a common method of NDT in extensive use. It is relatively simple to use for finding cracks that are open to the surface of a part. Liquid penetrants containing fluorescent dies are applied to the surface of a clean part to effectively wet the surface and to form a continuous and reasonably uniform coating. During this process the penetrant migrates into cavities that are open to the surface. After setting long enough, the excess penetrant is removed from the surface, and a developer is applied. When the die is removed from the surface, the penetrant remaining in the cracks will fluoresce under black light. Die penetrants do not need to put a magnetizing current through them, and they can detect cracks less than 0.125 mm (0.005 in.) in depth. There are several forms of this type of inspection, including the application of powders and developers, water-washable systems, post-emulsifiable systems, and solvent-removable systems. The quality of penetrant inspection is highly dependent on strict compliance with the inspection procedure and on careful cleaning of the parts so that any cracks are open and exposed before being subjected to the penetrant. The penetrant method is generally found to be more sensitive than x-radiography or ultrasonics for fine surface cracks.

The observation of indications on the developed surface after a penetrant-developer treatment is very dependent on the individual inspector and is, therefore, quite subjective. The true indications are considered to be of three general types (Ref 5):

- Continuous lines are observed generally as jagged lines; cold shuts appear as smooth, narrow, straight lines; forging laps appear as smooth wavy lines; and scratches tend to be shallow.
- Broken lines result from continuous lines becoming partially closed by working, such as grinding, peening, forging, or machining, and appear as a discontinuous line.
- Small, round holes are typically general porosity, gas holes, pinholes, or very large grains.

Ultrasonic inspection is a method of inspection in which a beam of high-frequency sounds is introduced into the material being inspected to detect surface and subsurface flaws. A transducer probe is placed against a test piece, with a coupling liquid used to provide a suitable sound path between transducer and test surface to increase the transmission of the ultrasonic pulse energy into the test piece. This is done to avoid having a layer of air between the probe and solid surface, which would transmit only a very weak beam. The sound waves travel through the material with some attendant loss of energy and are reflected at interfaces. The reflected beam is then detected and analyzed to define the presence and location of flaws. Previous heavy use of ultrasonic testing was used in areas such as inspection of casting, primary-mill products, forgings, flat-rolled products, rolled shapes, and welded joints. In addition, ultrasonic inspection techniques have been used for monitoring the initiation and presence of fatigue cracks.

Ultrasonics are being used to measure the case depth of induction-hardened parts such as axles, shafts, rollers, and bearings. An Ultrasonic Micro-Structural Analyzer (UMA), as shown in Fig. 12.3, uses a transducer to apply the sound-to-measure case depths from 1 to 20 mm (0.040 to 0.80 in.). The UMA measures the hardness depth by looking at the ultrasonic backscatter generated from the grain boundaries of the pearlitic core material and the transition zone between the martensitic case. The horizontal distance between the two lobes, the time of flight (TOF), is proportional to the case depth. Parts that have sharp transition zones produce the best accuracy of case depth measurement. Figure 12.4 shows the ultrasonic response from front-wheel-drive shafts induction hardened to different depths ranging from 3.2 to 5.3 mm (0.13 to 0.21 in.). As the case depth increases, the TOF is linear in respect to the case depth. The case-depth readings can be made in seconds with a direct reading of the case depth provided, and the readings are not sensitive to surface roughness, material properties, dimensional tolerances, small changes in temperature, or magnetic fields.

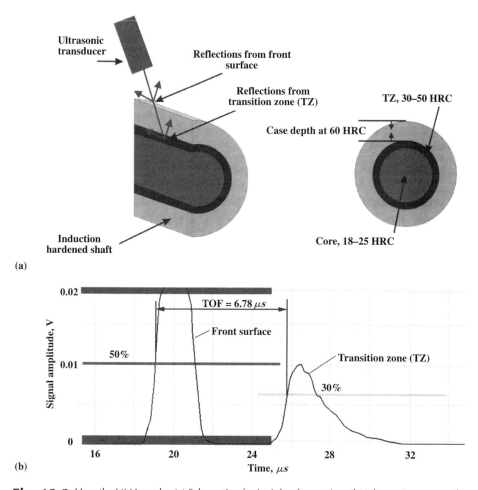

Fig. 12.3 How the UMA works. (a) Schematic of principle of operation. (b) Ultrasonic response for a hardened steel shaft. The time of flight (TOF) between the front surface and the transition zone (TZ) lobes is measured and then converted to a case depth accurate within 5%. Source: Ref 6

Eddy-current instruments operate by measuring impedance changes in a test coil. The instruments use wire-wound coils (Fig. 12.5) placed around or in the part to be tested. Any change in the test part, such as electrical conductivity and magnetic permeability, affects the coil impedance. These changes can be measured and monitored in the instrumentation with inspection results used mainly to make "go, no-go" decisions. Most eddy-current testing processes are designed for application on a particular part for inspection, such as checking cracks within 1 cm (0.4 in.) of the surface, surface hardness, and case depth.

Similar to induction heating, the depth of eddy-current penetration is determined by frequency. Current testing equipment uses computer-based multi-frequency, in-line test stations for the monitoring of case depth and hardness for various induction-hardened parts. Table 12.7 shows some of these applications of eddy-current testing. Table 12.8

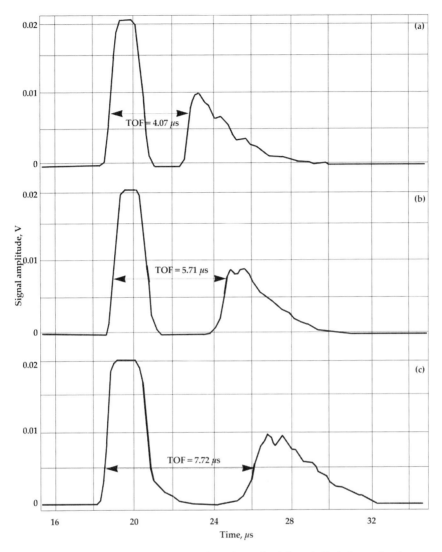

Fig. 12.4 Ultrasonic responses from front-wheel-drive half shafts made of Society of Automotive Engineers (SAE) 1050 steel and induction hardened to different depths: (a) 3.2 mm (0.13 in.), (b) 4.3 mm (0.17 in.), and (c) 5.3 mm (0.21 in.). As case depth increases, the transition zone lobe moves to the right; its position is linear with respect to case depth due to the relatively constant velocity of the sound waves. Source: Ref 6

shows some of the possible hardening errors that the equipment is capable of monitoring during induction heating. In-line processing requires the development of application-specific, eddy-current coils and extensive pilot testing of parts to develop the computerized test program. One of the main problems for use is that while nonconforming parts are identified, false trips occur because of the difficulty of developing programs that are accurate enough to handle all material and process conditions.

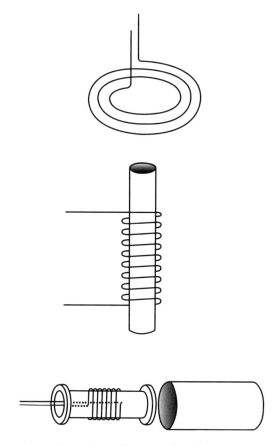

Fig. 12.5 Inspection coils configurations for eddy-current testing. The inspection coils have different configurations depending on the specimen shape. (a) A flat surface is normally examined by a flat pancake-type coil. (b) A cylindrical specimen is examined using an encircling coil. (c) The interior of a tube can be examined by an inside, inserted, or bobbin coil. Source: Ref 3

Radiographic inspection is the technique of obtaining a shadow image of a part using penetration radiation. The contrast in a radiograph is a result of different degrees of absorption of x-rays in the part depending on variations in part thickness, chemical constituents, nonuniform densities, flaws, discontinuities, or to scattering processes with the part. Several closely related techniques are used for different purposes. x-ray and some-

Table 12.7 Application examples of material testing by means of eddy current testing

Component	Parameters to be tested
Bearing parts	Case depth, structure and hardness pattern
Axle components	Case depth, structure and hardness pattern
Pinion pins and axles	Case depth, structure and hardness pattern
Linear guide components	Case depth, structure and grinder burn
Steering racks and the like	Case depth, structure and hardness pattern

Source: Ref 5

Table 12.8 Possible hardening errors during induction hardening

Which parameter was incorrect?	In which way was it incorrect?	What are the effects of this?
Austenitizing temperature	Too high	Overhardening, incorrect structure martensite + residual austenite
	Too low	Underhardening, incorrect structure martensite + bainite + ferrite
Austenitizing time	Too long	Overhardening, case too high, incorrect structure martensite + residual austenite
	Too short	Underhardening, shallow case, incorrect structure martensite + bainite + ferrite
Quenching	Too fast	Incorrect structure martensite + residual austenite
	Too slow	Incorrect structure martensite + bainite + ferrite
	Formation of vapor bubbles	Soft spots not defined
Annealing temperature	Too high	Hardness too low
	Too low	Hardness too high
Annealing time	Too short	Hardness too low
	Too long	Hardness to high
Rate of feed	Too slow	Shallow case, misplaced case, Austenitizing time too short
	Too fast	Case too high, misplaced case, austenitizing time too long
Damaged inductor	Undefined	Undefined
Malpositioning	Undefined	Unsymmetrical hardening pattern, overheating, melting

Source: Ref 5

times neutron diffraction are used on induction heat treated parts for the determination of residual stresses, as discussed in the Chapter "Heat Treating Basics" in this book. X-ray inspection can also be used for the determination of internal flaws, such as voids and cracks, but the orientation of the part being tested is very important in finding flaws such as cracks.

Inspection for Induction

Inspection plans should be completed for each induction-hardening process that incorporates all appropriate standards, and specifies what is to be inspected, the method of inspection, and the frequency of inspection. At first article inspection, all heat treating requirements need to be inspected. Once in production it is common for hardness to be checked only once per hour, and case depth once per shift. Inspection frequency is application specific, and testing programs should be designed to meet the individual part requirements. The amount of statistical process control (SPC) used for in-line testing will affect the requirements of off-line sampling frequency. Personnel using the test equipment need to be trained and tested in the operation of all testing equipment. Finally, proper workpiece preparation for testing is necessary for obtaining valid test results, with testing instructions provided to the personnel performing inspection.

Inspection of workpieces that are induction hardened requires the use of standard preparation techniques. This section presents tried and proven techniques for preparing and inspecting induction-hardened parts. This includes preparation for inspecting, hardness testing, and case-depth measurement.

Workpiece Preparation for Surface Testing. The surface of the workpiece must be free from oil, scale, and pits. According to the standards, test parts should be smooth and free of machining marks; however, the Rockwell C scale is not affected by shallow machining grooves, while the A and 15N scales are more affected. Standardized test blocks traceable to National Institute of Standards and Technology (NIST) must be used to calibrate Rockwell testers. Therefore, the most accurate testing is done with parallel surfaces. If the part is round, a v-block can be used, and special anvils can be used to hold the part to be checked directly under the indentor.

Removal of the quenchant and any other coatings or deposits including scale is necessary to prepare an induction-hardened part for inspection. The quenchant and any machining oils can be removed by immersion in an appropriate solvent. Glass bead is a quick method for scale removal that will not only clean, but will also produce etching or polishing that outlines the hardness pattern. If glass bead is not available, wire brushing, steel wool, or Scotch-Brite (3m Co., St. Paul, MN) can be used. At this point the part should be viewed for any nonconformances such as quench cracks, melts, or pits. Next, the bottom surface of the part that is under the area to be checked should be inspected to make certain it is flat and free of burrs. Parts that do not have parallel surfaces need to have the bottom ground parallel to the surface to be checked. Parts that are round can be checked on v-blocks with round correction added to the reading. Long workpieces may need to have the ends held by a steady rest for accurate, flat location of the workpiece on the anvil.

Decarburization as a Source of Error. The lighter test scales produce softer hardness readings with decarb than the heavier scales. Therefore, when two different test scales are used, and the hardness between the two scales does not correlate, decarb becomes a possibility. Decarb is not generally found over 0.25 mm (0.010 in.) on cold drawn stock, and sanding or grinding easily removes most decarb. When the stock removal is not complete, or when there is slight decarburization, carbon depletion from the surface may be only partial. Carbon gradients may not show decarburization through etching. However, microhardness testing will verify the hardness increase that indicates carbon depletion. An increase in surface hardness after slight stock removal is also an indication of surface decarburization.

Minimum Case-Depth Verification by Surface Hardness. Occasionally workpieces have minimum case depth specified only. An example would be where a 0.5 mm (0.020 in.) case depth is specified. When a Rockwell test scale is used that will support the indent, the minimum case depth could be certified without checking. An example would be obtaining an HRC 62 reading on the surface of a part. This is a clear indication that the case depth must be at least 0.7 mm (0.028 in.), and the case depth can be certified to 0.7 mm (0.028 in.) minimum. However, most of the time the case-depth range is specified so that the workpieces have to be cross-sectioned for case depth determination.

Cross-sectioning requires the use of a water-cooled metallurgical cut-off saw. (Water jet cut-off has also been used.) Because hardened parts are being cut, soft-bonded cut-off wheels must be used. Parts must be clamped so that they can be cut without causing binding of the cut-off wheel during cutting. In addition, parts with thin cross sections need to be mounted vertically so that the side of the part opposite the cutting wheel is adequately cooled during cutting. When effective case depth is measured (by hardness), either the first cut must be parallel to the bottom, the second cut must be parallel to the first cut, or the cut section must be parallel to the ground. Any time a workpiece is cut, there is a possibility of producing wheel burns (grinding burns). These are localized areas in which the martensite is heated during cut-off to a high enough temperature for tempering. The wheel burns will produce low hardness readings, and they may be detected by etching as described later in this Chapter.

Etching for Total Case-Depth Measurement. Figure 12.6 lists the procedures for mixing etchants. Lab procedures should be established for the safe mixing and handling of these etchants, including the wearing of protective eyeglasses when mixing. The commonly used etchants in induction heat treating are weak solutions of nitric and hydrochloric acids. Nital, a mixture of nitric acid and alcohol, is used most commonly for case-depth measurement with small concentrations ranging from 2 to 5% safely used. A weak solution of 5% hydrochloric acid and alcohol is used to define wheel burns. The process of etching for burns requires first overetching with nital until the etched area is as dark as possible. Then rinse with water, followed by alcohol, and blow dry. Finally, etch a second time with 5% hydrochloric acid. Any burned areas caused by tempering will remain dark, with the martensitic areas white.

After it has been determined there is no wheel burn (the double-etching test can be eliminated if you have determined from previous experience that there is no wheel burning occurring during the cutting of a particular part), polish the surface and etch with nital. After the etching process outlines the hardened case and provides definition of the core microstructure, the case depth can be optically measured.

> **Etchants:**
> Two principal etchants are used for etching in induction heat treating:
>   Nital (5%): 5% nitric acid, 95% alcohol
>   Mild Hydrochloric (5%): 5% hydrochloric acid, 95% alcohol
> **Procedure:**
> A protective rubber apron is recommended, and protective glasses are mandatory. Laboratory grades of alcohol can be purchased with ethanol (CH_3CH_2OH) recommended. Methyl alcohol, (CH_3OH) should not be used because of potential toxicity. Nitric acid and hydrochloric acid can be purchased in gallon or smaller bottles. The areas where the etchants are mixed should be well ventilated.
>
> Acid is always added to the alcohol. Try to keep the acid off of your hands, and wash any etchant off the skin as soon as possible. If any irritations occur after use, the affected individuals should wear rubber gloves when etching. The etchants can be placed in plastic or glass bottles. Etchants are applied onto the top of the surface to be etched by whatever technique seems best with test specimens immersed, sprayed, doused, or swabbed. (Cotton swabs may be used.) When the etching is complete, the surface is first cleaned with water, then alcohol, and finally blown dry. Once dry, the test specimen must be examined to make certain that there are no water spots or etching stains. Thereafter, the etched surface must not be touched. If the specimens need to be stored, plastic containers or cotton taped over the top is recommended.

Fig. 12.6 Laboratory procedures for etching in induction heat treating. Source: Ref 7

Measuring Effective Case Depth. Effective case depth is the case depth as defined by a given hardness reading, most commonly HRC 50 or the equivalent. Individual specifications by different manufacturers sometime provide their own definition of the hardness that is to be used. On parts for which the case depth exceeds 1.5 mm (0.060 in.), the easiest method of testing, if permitted, is to cut a parallel section. Then take HRC readings at the surface, extending toward the center in a spiral pattern, and the effective case depth at which HRC 50 is reached. Lighter cases such as the HR15N scale can be used with shallower cases as long as the hardness indent is 2.5 times the indenter diameter from the surface of the cut workpiece. Some workpieces are hard to measure, and if a parallel cut cannot be made, the workpieces can be mounted in epoxy for testing by microhardness.

Crack Detection. Cracks can often be seen on the workpieces after cleaning. The very first step in inspection, as previously mentioned, is to look for cracks. However, NDT methods such as magnafluxing may be needed to detect fine cracks. When the workpiece is inspected for cracks, it should be cooled to room temperature in the same manner as it will cool in production. (Setup personnel have a tendency to take a warm or hot workpiece out of the coil and then water cool so that testing can be done.) As discussed in the discussion of NDT, various types of NDT testing have been used to test for conformance to specifications including magnaflux, dye penetrant, eddy-current, and ultrasonic testing. Magnaflux and dye penetrant were used off-line to check for cracks, while eddy-current and ultrasonic testing can be used either on-line or off-line.

The *ASM Handbook* on mechanical testing (Ref 4) has very good detail on hardness testing and a detailed list of the ASTM standards for hardness testing. The ASM book *Nondestructive Testing* (Ref 5) outlines in detail ASTM and other standards relating to NDT methods.

REFERENCES

1. *Nondestructive Inspection and Quality Control*, Vol 11, *Metals Handbook*, 8th ed., American Society for Metals, 1976
2. *ASTM Standards*, ASTM
3. "Nondestructive Testing," Vol 03.03, Section 3, *Annual Book of ASTM Standards*, ASTM
4. *Mechanical Testing and Evaluation*, Vol 8, *ASM Handbook*, ASM International, 2000
5. L. Cartz, *Nondestructive Testing*, ASM International, 1995
6. M. Whalen et al., Ultrasonic Measurement of Case Depth, *Heat Treat. Prog.*, June 2000
7. R.E. Haimbaugh, Induction Heat Treating Corp., personal research

CHAPTER **13**

Nonconforming Product and Process Problems

THIS CHAPTER is designed to provide a step-by-step problem-solving sequence to help analyze the causes of nonconforming workpieces and to provide potential solutions. Workpieces can be out of tolerance for not meeting a hardness specification, case-depth specification, pattern length, or having a defect such as melting or a quench crack. Recommendations for inspection and testing are discussed in the Chapter "Standards and Inspection" in this book, along with methods for valid and reliable test results. Any time workpieces are checked with hardness testers and are found not to be in conformance with the required hardness, specimen preparation and testing methods should be reviewed for conformance to ASTM standards. Some of the more prominent situations that are encountered during testing and give bad test readings follow.

Improper Testing Procedures

The Hardness Test is Penetrating through the Case. ASTM Standard E 18-88, with Table 12.1, shows the maximum depth of case that a given hardness scale can correctly check. Sometimes the hardness specifications shown on prints are used for the testing but are not the scale that should correctly be used according to the ASTM standards. In such a case, the standards permit the use of the appropriate lighter scale, with the reading then correlated through the ASTM table (Table 12.2). The recording of the test results should include the scale readings tested and the appropriate scale correlation.

Round Correction Is Not Being Added to the Test Reading. Rounds greater than 6.25 mm (1/4 in.) in diameter need to have a round correction factor added (Table 12.2).

Specimen Preparation. The specimen surface is not smooth, and the surface is not perpendicular to the diamond indenter.

The Rockwell Diamond Indenter Is Chipped or the Rockwell is not in calibration.

Workpiece Has Defects As-Received

The induction heat treating process can cause defects that were originally in the parts to become more apparent. Examples of these are seams, porosity (in castings), cold shuts and cracks produced by cold heading, and other casting and forming defects. Seams and laps are discussed later in this Chapter under "Cracking." Anytime something unusual is seen, the workpiece should be carefully examined to make certain that the condition is not caused by a defect in the material. Most of the time the nonconformance will occur when one or more of the heat treatment specifications are not be met, as in the case of low surface hardness, wrong case depth, cracks, surface defects, and distortion. High surface hardness is not considered a problem because proper procedures call for tempering if the workpiece is over the specified hardness. The balance of this Chapter will focus on the typical nonconformances found during induction heat treating and possible remedies. Inspection normally starts with first article inspection followed by a system for periodic testing of the workpieces for conformance to specification. First article inspection will normally reveal more of the problems that can cause nonconformance than will appear during production, because some of the preexisting conditions that can cause the nonconformances are identified during first article inspection. The conditions that can cause nonconformance are discussed in the order of inspection frequency of occurrence. Low hardness probably is one of the largest areas of nonconformance.

Low surface hardness or soft spots are found on first article inspection. As previously outlined, the inspection techniques are verified for correctness and appropriateness. If the individual doing the testing does not see the workpiece when austenitized, another workpiece should be heated and quenched to visually verify that the austenitizing temperature was high enough and that the quenching looked good. Inspection techniques for determining soft spots are outlined in the Chapter "Standards and Inspection" in this book.

Decarburization. Check for decarburization (decarb) by the most appropriate method. Simple surface-material removal sometimes is the fastest method. Decarb exists if the carbon content of the surface is lower than the nominal carbon content of the workpiece. In some cases the surface decarb can be difficult to detect because the decarb layer is very thin or has a carbon content just under the average workpiece content, leading to slightly lower hardness just at the surface. This layer may quench to 100% martensite but of a lower hardness. In this case a slight grind or a microhardness test may be the only test that can determine that the decarb layer exists. Determine if further machining will be done. If so the decarb may not be detrimental to the use of the workpiece. Check the hardness below the decarb, and if the workpiece is in specification, note on the inspection record that decarb was present and that the hardness reading was taken below the decarb. Decarb does not always have a uniform distribution. Workpieces made from cold-drawn steel may be machined with more stock removed on one side of a bar than on the other. In this case the decarburized area would run longitudinally down one side of the workpiece. Also, cast or forged workpieces may not have the decarb totally removed in the machined areas.

Workpiece Not Quenched Fast Enough. If the overall surface hardness is low, the quenchant speed may need to be increased. Quenching speed depends on the quenchant being used, quench temperature, and quench impingement. If polymer is being used, decrease the concentration (this is most often easier and more controllable than trying to change the temperature). If oil is being used, increase the oil impingement or temperature. The Chapter "Tempering" in this book covers the selection and use of quenchants.

Workpiece Not Heated Hot Enough. Increasing the austenitizing temperature is a matter of increasing the power output of the power supply, increasing the heating time, or both. Usually when making the setup, the power is adjusted and set for the approximate time desired to heat the part. Then the heating time is used for fine adjustment of the cycle. The use of electronic timers and programmers makes adjustment of time cycles easy. If more power cannot be provided and adjusting the time does not help, the tuning of the power supply and heat station should first be examined to make certain that maximum power is being produced. If the tuning is correct and more power is needed, then the buss design used to attach the induction coil, as well as the induction coil design itself, should be reviewed. Better lead construction on the buss attachment and better coil design for coil efficiency may be needed.

Material Has Carbon Below Specifications. If the low hardness is not due to decarb, check to make certain that the material is not a low-carbon heat of steel. This can be done quickly by spark testing. As mentioned in the Chapter "Standards and Inspection" in this book, low-carbon steel produces very few sparks when held against a grinding wheel. Sometimes where there is decarburization, such as with parts that have been quenched and tempered in a neutral atmosphere, keeping the atmosphere slightly rich can make the difference in the ability to quench the surface hardness above specification.

Prior microstructures for induction heat treating need a fine, homogeneous distribution of the elements and phases (see the Chapter "Heat Treating Basics" in this book). Analysis of the microstructure is needed to define the steps necessary, if possible, to produce the fully martensitic microstructure. Examples of poor prior microstructures are workpieces that are found to be fully annealed, workpieces that have ferrite banding, workpieces that have segregation of some of the elements such as manganese, and workpieces that have a mixture of coarse and fine grains. All require higher austenitizing temperatures.

Fully annealed microstructures are difficult to fully austenitize and to diffuse the carbon. For example, 4150 steel annealed may only oil quench to 53 HRC, while quenched and tempered it may quench to 58 HRC or higher.

Ferrite banding is described in the Chapter "Heat Treating Basics" in this book in the discussion of undesirable microstructures. If the banding is severe enough, complete austenization with uniform diffusion of carbon may not be possible. Problems also occur with mixed heats of steel where the first article workpiece does not have banding, but other heats with banding are mixed in later production. Workpieces with ferrite banding running at the same austenitizing temperature will quench to a lower hardness because the carbides are not fully dissolved and dispersed in the matrix. In addition, the manganese segregation in banded steels produces areas in the core that austenitize at a lower temperature than the remainder of the part. This can result in martensite formation in the core below the effective case depth. If possible, heat identification and separation should be implemented.

Coarse-grained microstructures can be produced when hot rolling produces nonuniform grain sizes around the diameter that affect the quenched surface hardness. Workpieces in which the soft area runs longitudinally in a slow spiral are an indication of this condition. When higher austenitizing temperatures are needed, the heating time is normally made longer. However, if the austenitizing temperature becomes excessive and produces other undesirable effects, as discussed later in this Chapter in the section "Overheating," then lower power and longer time cycles may need to be used. If the material cannot be purchased with a microstructure good enough for induction hardening, then the material needs to be normalized to produce a uniform, pearlitic microstructure prior to the induction heat treating practice.

The Carbide Particles Are Too Large for Solution. This is often the nature of carbide particles (cementite) in fully annealed steels. The best prior microstructure for induction heat treating is one that has a fine, uniform, carbide distribution. When the carbides are large and coarse, they may not completely dissolve when the part is austenitized. The workpiece reacts like a lower-carbon steel and does not produce the hardness expected. Sometimes this can be easily discovered by rehardening the same part. If it quenches harder with the second heat cycle, then the prior microstructure is not good enough. Further proof of this can be found by microscopic examination. Under the microscope the hardened area can be examined to make certain that there was fully austenitic solution. A 52100 steel, with its high carbon content, austenitizes very well if the steel is purchase annealed with the carbides in a homogeneous, granular form.

Vapor pocket formation (steam pocket) is produced by poor quenchant contact with the workpiece when first quenched. Water quenchants, particularly warm water, are prone to produce vapor pockets. Vapor pockets permit transition products to form, producing localized soft spots. The vapor pockets produce small soft spots that can be determined by etching, hardness tests, or microstructural analysis. The remedy is to redesign and improve the quenching. Cold water quenches much faster than warm water; however, water quench systems need to have an economical way to keep the water cold. Refrigerated chillers can be expensive to operate, and unless there is cheap availability of cold make-up water, it may be hard to keep the water temperature low enough to be effective. A slight addition of polymer up to about 1% in water will decrease vapor-pocket formation without slowing the quenching rate so that a warmer quenchant can be used. Other techniques such as better impingement designs and submerged quenching with impingement are used. The Appendix "Quench System Design" in this book discusses quench ring design.

Transformation Products Occur during Quenching, producing low hardness. Transformation products are an indication that the quenching speed is too slow. When workpieces do not quench hard enough and decarb is ruled out, the quenchant speed needs to be examined and increased, if possible. In general practice, the quenchant speed can be increased until the workpiece either cracks or has excessive distortion. If oil is being used, general practice is to change to a polymer quenchant (the Chapter "Quenching" in this book). Polymer quenchants have become widely used, with concentrations ranging from 1 to 28%. Concentrations of 1% quench with the approximate speed of water, while concentrations of 28% quench with the approximate speed of oil. In addition, there are special polymers that are reported to quench as fast as brine. Quenching curves, as supplied by the manufacturers, are based upon immersion quenching. Polymer quenchants, as used for induction, generally involve the use of spray quenching. Spraying polymer on the surface of a workpiece being quenched tends to decrease the thickness of the polymer layer. Therefore, it is not unusual to need to test the range of concentration that can be safely used for a particular workpiece. Too low of a concentration can produce quench cracks, while too high of a concentration can produce low surface hardness of case depth.

Removal from Quench Too Soon. Workpieces can self-temper if removed from the quenchant too soon. Examples of this are conveyorized fixtures in which the conveyor speed is too fast, or scanning where the quench is not applied for a long enough distance to fully quench the part. The remedy is to slow the conveyor or to modify the quench ring or pad to increase the quenching impingement and efficiency. Design factors that are to be changed include increasing the number of inlets, putting baffles inside the inlets if the output flow varies around the quench ring, changing the outlet hole size and locations, and possibly changing the impingement angle. Another design change that can be made is to add a secondary quench below the primary quench.

Barberpole. Spiral patterns caused by scale formation during quenching are often seen in workpieces that are scanned. When properly quenched, barberpole is a visual condition, and there is no effect on the hardened characteristics. However, if the quenching is not sufficient, two effects can occur. First, flakes of bainite may appear on the surface of the workpiece in a spiral pattern. These flakes can be seen in the microstructure, and if the workpiece is centerless ground, the resulting surface finish and tolerance can be affected. Second, barberpole can be the result of self-tempering of a martensitic area as it passes from one quench impingement area to another. When microscopically viewed, the workpiece may have a fully hardened case with a very shallow, tempered area spirally down the workpiece. In both cases the quenching must be improved.

Retained Austenite. Stainless steels, 52100 steel, and tool steels such as those in the D-series, when austenitized too hot, will have the quenched microstructure consisting of martensite, small carbide particles, and retained austenite. 52100 will normally have up to 10% retained austenite that is not visible and can be detected only through x-ray diffraction. When the retained austenite exceeds approximately 10%, some retained austenite starts to become visible in the microstructure. As the retained austenite increases, the quenched hardness will start to drop. An indication of overheating and retained austenite can be found by performing hardness tests from the unhardened area into the hardened area. The hardness first increases, then peaks, and then drops as the testing progresses into the overheated area. The peak in hardness is the point at which the austenitizing temperature was correct. The solution, of course, is to decrease the austenizing temperature.

Pattern Length Not in Conformance. If the workpieces are being scanned, then the distance for scanning needs to be adjusted. If the workpieces are being heated single shot and the pattern length is long enough but off center, the workpiece needs to be repositioned. If the coil length is not correct, then the coil needs to be recharacterized to produce a longer or shorter pattern as necessary.

Low hardness in a cut workpiece is a condition that is commonly produced by metallurgical cut-off wheels during cutting. Although the wheels are water cooled, if soft wheels are not used on hard materials, sufficient temperature can be produced during cutting to locally temper the parts. Any time low hardness is found after a workpiece is cut, the workpiece should be double etched as described in the Chapter "Standards and Inspection" in this book to check for burning. It may take some experimentation to find the best part orientation and clamping technique to prevent burns. Thin parts give particular problems, and it is best to cut thin sections standing vertical to prevent burns.

Shallow case depth is normally an indication that a longer heat cycle is needed. If a longer heat cycle does not produce a deeper case, then more detailed review is needed. Hardenability curves as shown in the Appendix "Hardenability Curves" in this book are based on a furnace-austenitized part and a fixed water quench. The furnace austenitizing done for the hardenability curve negates any effect of prior microstructure because there is full time for austenization. However, with induction hardening, hardenability is a function of the chemistry of the part, the prior microstructure, and the rate of quenching.

Hardenability is discussed in the Chapter "Heat Treating Basics" in this book. If a part does not produce a deep enough hardness with a longer heat cycle while appearing to have a large hardness transition zone, verify that the material is right and that the grade of material has the hardenability needed. Then check the quenchant. If polymer is being used, add water to the quenchant to increase the quenching speed. To increase quenching efficiency, the quenchant impingement can also be changed. Finally, if quenching is optimal, verify that the prior microstructure is not having an effect. Sometimes this can be done by simply reheating the same workpiece two or three times and seeing if the surface hardness and case depth both increase. (If the prior microstructure is having an influence, it should also produce a lower surface hardness.) The sequence as mentioned considers that the power supply frequency was fixed. If the power supply has variable frequency, lower the frequency. This will produce a deeper heating effect. Make certain that the frequency selection is capable of producing the case depth requirements as discussed in the Chapter "Theory of Heating by Induction" in this book.

That the microstructure is not 100% martensite can be an indication that the workpiece was not austenitized at a high enough temperature, or it was not quenched fast enough. However, as noted, the prior microstructure has a significant influence on the ability to produce a fully martensitic microstructure. The influence of the prior microstructure, as discussed previously in the section "Low Surface Hardness" in this Chapter has the same impact and resolution. Review of the microstructure is necessary to determine whether the workpiece has quenched to 100% martensite; the review of the microstructure will define the problem. Comparison with the core microstructure that is not affected by the austenitizing cycle will clarify whether the nonconformance is due to incomplete solution or due to quenching too slowly.

Case Depth Too Deep. This is one of the easy-to-remedy nonconformances. Decrease the heating time to lower the austenitizing temperature.

Cracks. Cracks are discussed in the Chapter "Decarburization and Defects" in this book. Sometimes a "ping" can be heard during the heating or quenching of a workpiece, providing a clear statement that a crack has occurred. Most of the time cracks are found on inspection after quenching. Then it is important to determine the nature of the crack and when it occurred.

Seams, Laps, or Defects. Inspect the workpieces before induction hardening in order to determine whether the crack is from a seam or defect that is present in the workpiece. Figure 13.1 shows a crack that can be seen near the end of a 35 mm (1.40 in.) diameter shaft after induction hardening. The "V" at the left hand side of the workpiece is of significance because it clearly indicates that the crack is present during heating. The "V" is the effect of the overheating that occurs when the current jumps across the crack in front of the main heating as the workpiece enters the coil. At this point of investigation the assumption can be made that the crack was either in the part or opened at the start of heating and propagated. Additional testing was done. Figure 13.2 shows a micrograph of the cross section from the same workpiece. A light line of surface decarburization can be seen running along the inside of the crack, a clear indication that the crack was produced in the workpiece during previous hot processing. Magnafluxing before induction hardening should have also defined the crack, as shown in the figure. Sometimes these cracks are not continuous

Fig. 13.1 Crack propagating longitudinally. View of induction heated area showing "flaring" of the heat pattern at the crack, and the crack propagating past the heated area. Approx. 1.5×. Source: Ref 1

in the bar stock and are present in only a small portion of the workpieces. Then the decision needs to be made as to whether to scrap the material or to run and to separate all cracked workpieces with appropriate inspection after hardening.

Stress cracking during heating occurs due to high, nonuniform residual stress distributions in the workpiece. Heating a workpiece without quenching (if possible) and inspecting for cracks is one procedure that may help. Other times, the distribution of these cracks in the workpieces, such as 1 to 5%, is so small that the cause is hard to determine. When stress cracks occur during heating, furnace stress relieving a workpiece from 455 to 538 °C (850 to 1100 °F) may reduce the residual stresses enough to eliminate the cracking. On large parts the technique of using a low-temperature preheat from the induction coil, followed by a high heat austenitizing cycle has been known to eliminate cracking after induction hardening.

Quench Cracks from Cooling Too Fast. When the entire cross section being hardened passes through martensitic transformation at the same time, the potential for quench cracking is reduced. If a workpiece has a quench crack, the quenchant may be too fast or too cold. The first standard technique employed is to use a slower speed quenchant. When polymer quenchants are being used, the polymer concentration should be increased. A second technique is to increase the quenchant temperature.

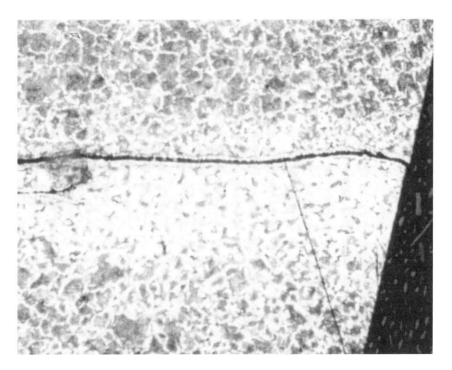

Fig. 13.2 Microphotograph of the crack showing decarburization on both edges. Nital etch; 130×. Source: Ref 1

When this is done, however, the quench tank needs to have heaters so that the quenchant is brought up to temperature before the start of production. Increasing the quench temperature of fast-quench oils up to the region of 78 °C (175 °F) actually increases the quenching speed. Some quench oils need to be heated up into the 120 °C (250 °F) region to slow the quenching speed. In applications for which timed spray quenches are used, the quenching time can be reduced so that the part is brought out hotter. If the *timing* is such that the part is 120 to 150 °C (250 to 300 °F), quenching can be stopped and the part can be removed hot. The workpiece is then air cooled with both the mass of the part and convection to the air doing the cooling. Full transformation to martensite occurs because the workpiece is in the middle of the martensite-transformation temperature region. Whatever temperature is used, the process must be capable of replication and producing the same as-quenched temperature throughout the daily production.

One of the main problems in removing workpieces hot is that unless mechanized, induction heater operators like to have the workpieces cool enough to handle without burning themselves. The other objection would be that where the workpiece is being subsequently induction tempered in-line, it would not be fully martensitic before the tempering operation. The potential then exists that after cooling from tempering, untempered martensite may be produced from incomplete austenitic transformation in the first cooling cycle.

In contrast to the technique of removing workpieces from quenching in the middle of the martensitic transformation range, there have been some workpieces for which intensive quenching techniques were used for very rapid quenching. The intention is to transform a case depth of martensite deep enough that the yield strength of the case is high enough to ensure that rupture does not occur when there is subsequent expansion of the workpiece caused by transformation or thermal expansion of the core.

Oxide inclusions can nucleate quench cracks (as seen in the Chapter "Decarburization and Defects" in this book). If a slower quenchant does not stop the cracking, then contact should be made with the steel supplier to see if the inclusion content can be reduced.

Workpiece Shape. The basic geometry of some workpieces can cause the initiation of quench cracks. For example, keyways may have sharp fillets at the bottom of the that can act as stress risers that initiate cracks. Keyways should always be machined after induction hardening, if possible. Sometimes variation of the frequency of the power supply, such as lowering the frequency to produce a deeper case, will help. Another technique is to change the orientation of the coil so that it is longitudinal to the keyway, producing less heat at the edges.

Lower the Carbon Content of the Workpiece. Some material specifications have excessive carbon. When the as-quenched hardness is considerably higher than the specified hardness range, lowering the carbon content will decrease the potential for quench cracks.

Prompt Tempering. Prompt tempering is always suggested to minimize quench cracking after the parts have been removed from the power supply.

Overheating Resulting in Pitted or Deformed Surface. As the austenitizing temperature is increased in air, heavier scale forms on the workpiece being heated. When workpieces are heated too hot, the scale formation can result in an actual pitting of the surface. Unless there are final metal-removal operations, pitting may be detrimental to the point that the final desired surface finish is not produced. Workpieces with holes, keyways, or changes in cross section should always be examined for melting in these areas. Also, the parts may be overheated to the point that actual sparking and melting of the surface can be seen. Finally, grey cast irons have a phase that starts to melt at about 903 °C (1700 °F). The melting is seen as an actual flow or fusion of the part. Overheating of the workpiece is an indication that time or power needs to be reduced, or the set up needs to be changed.

Too-Large Martensitic Grain Size. Martensitic grain size can be hard to define by microscopy. The workpieces must be tempered at a high enough temperature so that when the martensite is etched, the contrast is dark enough that the martensite packets and their orientation produce a cluster effect that will show the prior austenitic grain size. The specification should define the magnification to be used so that there is agreement in grain size definition. If the martensitic grain size is too large, the austenitizing temperature must be reduced.

Excessive Distortion. Reduction of distortion moves more into a trial-and-error process. First the distortion must be clearly defined on print tolerances. The Chapter "Heat Treating Basics" in this book discusses distortion. ASM Handbook, (Ref 2), provides a thorough discussion of distortion and design. Workpieces need to be designed so that the potential for distortion is reduced. Then the impact of items such as holding nests or fixtures, the quenchant used, timing of quenchant application, including primary and secondary quenchants, and the potential for machining to final size after induction hardening must be reviewed and tested. Sometimes if distortion is constant from workpiece to workpiece, the machining tolerances before induction hardening can be set so that the final net shape desired after hardening is produced.

Workpieces Are Not Heating the Same (hardness, case depth, pattern length, etc.). Conditions occur under which all of the previous potential problems have been reviewed and have been found to be in control or in proper range, but inspection is finding nonconforming workpieces. The following are physical conditions that occur outside the fixturing and materials:

- Variation in line voltage is causing the power supply output to change. In times of high power consumption it is not unusual to have fluctuation and variation of the incoming line voltage. The utilities are commonly allowed plus or minus 10% on a 480 volt supply. The manufacturers of power supplies design the power supplies to be able to produce constant output power over some variation in line voltage. The common type of output regulation is constant power, constant voltage, or constant current. However, particularly if tuned and operating at full output power, not all power supplies will produce constant output over severe line-voltage fluctuations. If the output power is not constant for a given heat-cycle time, the part being heated will not heat to the same temperature. The use of coil signature monitors with process characterization will indicate whether a part is heating to the identical desired parameters during each cycle. Some purchasers of new equipment specify that the power supply must operate at a given percent, such as 50 or 75%, of the rated output to give a wider window of operating power control.
- The workpieces are not locating accurately from heat cycle to heat cycle. All workpieces heated on a setup must be located in the same position in the induction coil. While the lower frequencies are more forgiving of eccentric rotation, the locations for the heat on (and, if scanning, the heat off) must be the same. It is highly recommended that workpieces have machined surfaces with consistent, accurate dimensioning to the surface to be heated. The workpieces to be heated must be machined the same. On high volume production, sensors should be present to make certain the workpieces are correctly seated on the holding mechanism. The sensors should also monitor that all mechanical operations have occurred in the correct sequence. Finally, all workpieces must be free from chips and burrs because these can interfere with proper seating.
- The timing cycle is not accurate enough. The shorter the heat cycle, the higher the degree of timer setting that is necessary. For heat cycles of less than 0.5 s, the timer should have settings to 0.001 s. For heat cycles between 0.5 and 2 s, the timer setting should be to 0.01 s, and for heat cycles greater than 2 s, timer settings to 0.10 s are generally sufficient for most applications. Note that on high power, single-shot applications, higher accuracy might be desirable. Good design would be to have the timer accuracy higher than is needed.

- The concentration material in the coil has deteriorated. Many of the specialized induction coils today use flux concentrators. While there are varying types of materials used, the common factor among the different materials is that degradation occurs. After a point, the degradation will cause the coil efficiency and flux concentration to change to the point that the parts being induction heated will no longer quench into specifications. The coils need to be examined, sometimes on an hourly basis, until some useful estimate of minimum coil life can be determined. Maintenance cycles can be established for the rebuilding of these coils.
- The coil is moving. Induction coils can vibrate and flex due to the electromagnetic fields established and due to the heating of the coil itself because of the high frequency alternating current (ac) current. Rigid design is important; lower frequencies tend to produce higher degrees of vibration. Techniques to improve the coil rigidity range from the use of studs on multiturn coils to machined coils with better cooling passages. It is important to cool the coil so that heating does not occur. Booster pumps can be placed on the inlet side of the coil to increase the water flow through the coil.
- There are mixed heats of material. Whenever possible in production, it is desirable to keep parts made from different heats of steel in the same production lot. The heats should be identified and kept separate. When the steel is purchased in smaller than heat quantities, there may be no guarantee that mixed heats are not being purchased. When there are different heating results and mixed heats are suspected, sometimes the difference in the two heats can be seen by simply checking the core hardness. Spectroscopic examination can be used for accurate chemical analysis.
- There is internal heating of components in the power supply or heat station. Restrictions or debris in the cooling lines of the internal components can cause overheating and change in the output of the power supply. Infrared sensors have been used, and any unusual oxidation or signs of overheating can indicate restrictions.
- There is oxidation in the electrical connections on the coil or output power transmission buss. Occasionally, heating a component that carries the high-frequency current from the power supply to the induction coil will cause a bolt to loosen, with the result that either the bolt starts to carry the current and overheat or the electrical connections themselves will oxidize. The electrical resistance is changed, and the output power to the induction coil will change. During maintenance, inspect for any signs of discoloration in the buss, the coil, or bolts.
- There is a short circuit in the output transformer. Radio frequency output transformers are very susceptible to failure due to internal arcing on the primary side. Usually the electrical short will be violent and sudden and will cause a power supply overload fault. However, there

are times when the arcs start out as a small electrical leakage and are not obvious. Arcing may occur at the induction coil, and the power supply may have an overload trip with no visible reason. Sometimes the power supply will reset and operate for a period of time before faulting again. The lower-output voltage solid-state power supplies can also have output transformers short. This will normally produce an overload or trip.

REFERENCES

1. Ajax Magnathermic, unpublished data
2. Tocco Inc., 30100 Stephenson Hwy, Madison Heights, MI 48071-1677

CHAPTER **14**

Quality Control

EFFECTIVE SYSTEMS of quality control/quality assurance are essential for modern day heat treating practices. The essentials of an effective control program include:

- An independent quality assurance department
- Standards of quality that reflect customer needs
- Written procedures covering all phases of the heat treating process, starting with prototype qualification through inspection approval for shipment
- Process control documentation
- Methods of maintaining part identification through heat treating and keeping written inspection records
- Inspection procedures including frequency sampling, and identification and segregation procedures for nonconforming products
- Schedules for calibration of test equipment
- Schedules and procedures for record retention
- Identification of training needed and implementation of training programs
- Systems for control of documentation including review and distribution
- Periodic audits

Formalized Quality Control

Formalized programs such as *International Organization for Standardization* (ISO) and Quality System (QS) containing all the elements in the preceding list plus more are now in place worldwide. These programs require that businesses are established and operated in a manner that satisfies specific requirements. Independent certified trainers and auditors have been established to help businesses operate according to the principles of the programs. The ISO certification requirements are based upon the

European and now worldwide standards. Quality System programs are based on U.S. automotive standards. The QS standards are more stringent than ISO. This Chapter deals with the essence of quality system and will also present methods for design of quality induction heat treating systems.

Formalized quality control is required by the automotive industry with the goal of developing fundamental quality systems that provide for continuous improvement. Defect prevention is emphasized, along with a reduction of variation and waste in the supply chain. Quality System 9000, ISO/TS 16949, or new versions of ISO 9000:2000 as established define the fundamental QS expectations. As of the publishing of this book, registration of Tier 1 suppliers is mandatory, and all Tier 2 suppliers must be registered before Jan. 1, 2003. As this book is being written, ISO 9002 and 9003 are being replaced with a new quality standard, ISO 9000:2000. In its new form, the standard involves three areas:

- Fundamentals and vocabulary
- Requirements
- Guidance for performance improvement

ISO 9000:2000 will underline the concepts and approaches for the ISO 9000:2000 family and will provide definitions for the new vocabulary. The purpose of ISO 9000:2000 is to be the support mechanism for the interpretation of ISO 9001:2000.

ISO 9000:2001 has the same role as ISO:1994 with the content and organization revised. ISO 9001:1994 involved 20 elements, while ISO 9001:2000 consists of four major areas:

- *Section 5:* management responsibility
- *Section 6:* resource management
- *Section 7:* product and/or service realization
- *Section 8:* measurement, analysis, and improvement

The new organization is intended to bring ISO 9001 more in-line with ISO 14001 (environmental) standards and is consistent with ISO 9004's plan-do-check-act improvement cycle.

ISO 9004:2000 goes beyond the basic requirements specified in ISO 9001 and is intended to be a guide for companies wanting to improve the quality system after implementing ISO 9001. ISO 9004 is not a specification and cannot be used by third-party auditors for registration or certification assessment.

The 1994 standards had continuous improvement goals, while the new standards require improvement from audit to audit. Certifications for the ISO 1994 standards will expire within three years of the date of final publication of ISO 9001:2000.

The standards define the QS requirements and provide a detailed QS assessment (QSA) format. Included are advanced production quality planning and control plans. Of particular interest to heat treating is the production part approval process (PPAP) that is required for induction-hardened parts. Instructional manuals with procedures can be obtained from the Automotive Industry Action Group (AIAG), Ref 1.

Qualify System Assessment is the first step in formalized certification and is used in several different ways according to the needs of the customer and supplier:

- First-party assessment is self-assessment of one's own quality system, with certain minimum requirements and questions to be answered.
- Second-party assessment is assessment by the supplier of its customer's quality system, with certain minimum requirements and questions to be answered. This includes Tier 1 suppliers' assessment of all subcontractors.
- Third-party assessment is an assessment by an independent auditor/registrar. The assessment method is composed of three major phases: Phase I is a review that determines whether the quality manual and supporting documentation, as required, meet all requirements. Phase II is an on-site audit to determine the degree and effectiveness of the implementation of the quality system at the supplier's site. Phase III is an analysis and a report to provide the finding of the first two phases. Essentially at this point the auditor/customer will present the findings and decide whether the supplier is recommended/not recommended or has what is called a "variable score." Major nonconformances may be noted which represent an opinion that a total breakdown in the system will result in the failure of the quality system to assure controlled processes and products and that nonconforming products are likely to be shipped. Minor nonconformances are those that are not likely to result in the probable shipment of nonconforming product.

Essentially an overall evaluation of "recommended" will be given when the audit does not identify any nonconformance. An "open" status exists when either a major nonconformity is noted in the audit, or one or more minor nonconformances are noted in the audit. Open status can be converted to a recommended status within 90 days, or an otherwise agreed-to timeframe.

Production Part Approval Processes require the production run of one shift or less, of workpieces on the induction heat treating system that is to be used for production. During this production run, considerable documentation and testing must be done. On shipment, a Part Submission Warrant (Fig. 14.1) must be sent with the parts.

According to the standards, submission of PPAPs is always required before the first production shipment of product in the following situations:

- A specific new part is supplied to a specific customer.
- A discrepancy on a previously submitted part is being corrected.

Part Submission Warrant

Part Name ①_____ Part Number ② _____

Safety and/or
Government Regulation ☐ Yes ③ ☐ No   Engineering Drawing Change Level ④ _____ Dated _____

Additional Engineering Changes ⑤ _____ Dated _____

Shown on Drawing No. ⑥ _____ Purchase Order No. ⑦ _____ Weight ⑧ ____ kg

Checking Aid No. ⑨ _____ Engineering Change Level ⑩ _____ Dated _____

SUPPLIER MANUFACTURING INFORMATION   **SUBMISSION INFORMATION** ⑬

⑪
Supplier Name   Supplier Code

☐ Dimensional   ☐ Materials/Functional   ☐ Appearance

Customer Name/Division ⑭ _____

⑫
Street Address

Buyer/Buyer Code ⑮ _____

Application ⑯ _____
City/State/Postal Code

REASON FOR SUBMISSION ⑰
☐ Initial Submission
☐ Engineering Change(s)
☐ Tooling: Transfer, Replacement, Refurbishment, or additional
☐ Correction of Discrepancy

☐ Change to Optional Construction or Material
☐ Sub-Supplier or Material Source Change
☐ Change in Part Processing
☐ Parts Produced at Additional Location

☐ Other – please specify

REQUESTED SUBMISSION LEVEL (Check one) ⑱

☐ Level 1 – Warrant, Appearance Approval Report (for designated appearance items only).

☐ Level 2 – Warrant, Parts, Drawings, Inspection Results, Laboratory and Functional Results, Appearance Approval Report

☐ Level 3 – At Customer Location – Warrant, Parts, Drawings, Inspection Results, Laboratory and Functional Results, Appearance Approval Report, Process Capability Results, Capability Study, Process Control Plan, Gage Study, FMEA.

☐ Level 4 – Per Level 3, but without parts.

☐ Level 5 – At Supplier Location – Warrant, Parts, Drawings, Inspection Results, Laboratory and Functional Results, Appearance Approval Report, Process Capability Results, Capability Study, Process Control Plan, Gage Study, FMEA.

SUBMISSION RESULTS ⑲
The results for ☐ dimensional measurements ☐ material and functional tests ☐ appearance criteria ☐ statistical process package
These results meet all drawing and specification requirements: ☐ Yes ☐ NO (If "NO" – Explanation Required)
⑳
DECLARATION
I affirm that the samples represented by this warrant are representative of our parts and have been made to the applicable customer drawings and specifications and are made from specified materials on regular production tooling with no operations other than the regular production process. I have noted any deviations from this declaration below:

EXPLANATION/COMMENTS: ㉑ _____

Print Name _____ Title _____ Phone No. _____

Supplier Authorized Signature ㉒ _____ Date _____

―――――――― FOR CUSTOMER USE ONLY ――――――――
☐ Approved   ☐ Rejected   ☐ Other _____
Part Disposition
Customer Name _____ Customer Signature _____ Date _____

Fig. 14.1 Part submission warrant Source: Ref 1

Completion of the Warrant

PART INFORMATION

1. **Part Name** and 2. **Part Number:** Engineering released finished end item part name and number.
3. **Safety/Regulated Item:** "Yes" if so indicated on part drawing, otherwise "No."
4. **Engineering Change Level & Approval Date:** Show change level and date for submission.
5. **Additional Engineering Changes:** List all authorized engineering changes not yet incorporated on the drawing but which are incorporated in the part.
6. **Shown on Drawing Number:** The design record that specifies the part number being submitted.
7. **Purchase Order Number:** Enter this number as found on the purchase order.
8. **Part Weight:** Enter the actual weight in kilograms to three decimal places.
9. **Checking Aid No.** Enter the checking aid number, if one is used for dimensional inspection, and,
10. Its **Engineering Change Level** and **Approval Date.**

SUPPLIER MANUFACTURING INFORMATION

11. **Supplier Name & Supplier Code:** Show the code assigned *to the manufacturing location* on the purchase order.
12. **Supplier Manufacturing Address:** Show the complete address of the location *where the product was manufactured.*

SUBMISSION INFORMATION

13. **Submission type:** Check box(es) to indicate type of submission.
14. **Customer Name:** Show the corporate name and division or operations group.
15. **Buyer Name:** and Buyer Code: Enter the buyer's name and code.
16. **Application:** Enter the model year, vehicle name, or engine, transmission, etc.

REASON FOR SUBMISSION

17. Check the appropriate box. Add explanatory details in the "other" section.

REQUESTED SUBMISSION LEVEL

18. Identify the submission level requested by your customer.

SUBMISSION RESULTS

19. Check the appropriate boxes for dimensional, material tests, performance tests, appearance evaluation, and statistical data.
20. Check the appropriate box. If "no," enter explanation in "comments" below.

DECLARATION

21. **Comments:** Provide any explanatory details on the submission results; additional information may be attached as appropriate.
22. The responsible supplier official, after verifying that the results show conformance to all customer requirements and that all required documentation is available will sign the declaration and provide **Title** and **Phone Number.**

FOR CUSTOMER USE ONLY

Leave blank.

Fig. 14.1 (continued) Part submission warrant. Source: Ref 1

- The part was modified by an engineering change to specifications or materials. The submission of PPAPs may be waived for the following circumstances, (however all items in the PPAP file must be reviewed and updated):
  - A material other than what was used in the previously approved part is being used.
  - The production is from new or modified tools (except perishable tools), including additional or replacement tooling.
  - Production is following refurbishment or rearrangement of existing tooling or equipment.
  - Production is following any change in the process or method of manufacturing.
  - Production from tooling and equipment has been transferred to a different plant location or from an additional plant location.
  - Source for subcontracted parts, materials, or services (such as heat-treating) are being changed.
  - Product is being re-released after the tooling has been inactive for volume production for 12 months or more.
  - A customer requests to suspend shipment due to a supplies quality concern.

Five different submission levels are on the warrant, with the common submission level for induction heat treating being the requirements of level 3. The usual requirements for induction-hardened parts at the minimum are a completed copy of the warrant with completed workpieces, inspection results, process capability results, process control plan, gage study as applicable, and a failure mode and effects analysis (FMEA).

Control plans provide a written summary description of the systems used in minimizing process and product variation, and they are an integral part of an overall quality process. The required actions for each phase of the heat treating process are described to ensure that all process outputs will be in a state of control. Because processes have the expectancy of continuous improvement, a control plan reflects a strategy that is responsive to changing process conditions. Controls plans are updated as measurement systems and control methods are evaluated and updated. Figure 14.2 illustrates a control plan for a specific workpiece that was induction hardened. Current instructional manuals should be purchased before attempting to complete a control plan or a FMEA.

The FMEA is a systems failure mode and effects analysis, which is a disciplined analytical technique that assesses and assures, to the extent possible, that potential failure modes and their associated causes/mechanisms have been addressed. Through use of the systems, the degree of risk can be determined, along with plans of recommended action for reduction

Quality Control / 255

INDUCTION HEAT TREATING CORPORATION CONTROL PLAN

Customer: XXX Automotive
Part Number: XOOOOOOX
IHT Approval: _____
QC contact: _____
Production contact: _____

Original FEMA Date: _____
Revised IEMA Date: _____
Revision No.: _____

Process	Machine/device	Characteristics		Special char. class	Product spec. tolerance	Methods			Control method	Reaction plan
		Product	Process			Evaluation/ measurement technique	Sample size	Frequency		
First Piece Inspection	Setup	Surface hardness	...		HRC 59.5/66.0	Microhardness tester	1	Each setup	Cut parallel slice, polish through 600 paper, check hardness on contact angle at 1.2 mm deep per sketch.	Correct process and repeat inspection
	Setup	Case depth	...		2.7/3/7 mm tester	Microhardness	1	Each setup	Cut parallel slice, polish through 600 paper, check hardness on contact angle per sketch	Correct process and repeat inspection
	Setup	Run in/run out Pattern	...		6.5/10.5 mm	Digital caliper	1	Each setup	Cut parallel slice, polish through 600 paper, macroetch with nital	Correct process and repeat inspection
	Setup	Microstructure	...		100% martensite	Microscope 400×	1	Each setup	Cut parallel slice, polish through alumina, etch with nital, check per sketch	Correct process and repeat inspection
Hardening	Machine	Surface hardness	...		HRC 59.5/66.0	Microhardness tester	1	Each hour	Cut parallel slice, polish through 600 paper, check hardness on contact angle at 1.2 mm deep per sketch.	Correct process and repeat inspection
	Machine	Case depth	...		2.7/3/7 mm	Microhardness tester	1	Each hour	Cut parallel slice, polish through 600 paper, check hardness on contact angle per sketch	Correct process and repeat inspection
	Machine	Run in/run out pattern	...		6.5/10.5 mm	Digital caliper	1	Each hour	Cut parallel slice, polish through 600 paper, macroetch with nital	Correct process and repeat inspection
Quench	Operator		Quench concentration		8% +/− 2%	Refractometer	1	Start up and weekly	Measure concentration & record	Rectify as needed results on inspection log
	Machine		Quench temperature		90 F +5/− 15 F	Controller	1	Every 4 h	Read & record on inspection log	Rectify as needed
Temper	Machine	Surface hardness	...		350F +/− 5 F	Microhardness tester	1	Every load	Place slice in separate container in with load	If hardness is high, segregate and call customer to authorize higher tempering temperature.

Fig. 14.2 Control plan. Source: Ref 2

INDUCTION HEAT TREATING CORP.

Process FEMA

Customer: XYZ Automotive **Power Supply: 260-1**

Part Numer: Hub XX

IHT Approval: ____________

Original FEMA Date:

Revised FEMA Date:

Revision No.:

Characteristic ID/Name	Potential Failure Mode	Potential Effect(s) of Failure	Sev	Class	Potential Cause(s) of Failure	Occ	Current Process Controls	Det	RPN	Recommended Action(s)	Responsibility & Target Completion Date	Action Results: Actions Taken	Action Results: Sev	Action Results: Occ	Action Results: Det	Action Results: RPN
Storage																
Oil free hubs	Quench contamination	Reduction of Pattern	2		Too much oil by CNC	2	Wash if necessary	1	4	None						
Load Fixture																
Hand Load Parts on Hub	Load Hub Wrong Way	Improper Case Pattern	6		Operator Error	1	Cause Control: Standard Operator Instruction	1	6	None						
Austenitizing																
Heat Hub to Austenitizing Temp.	Power too High	Case too deep	6		Wrong Set up	1	Cause Control: Set up Instruction	1	6	Note						
	Power too Low	Shallow case, soft Hub	6		Coil Deterioration	1	Cause Control: Standard Operator Instructions, Inspection Plans	1	6	None						
	Time too Long	Case too deep	6		Wrong Set up	1	Cause Control: Set up Instruction, Inspection Plan Characteristic Control: PLC Controls, Part Position, Power Supply Regulation, Quench Control,	1	6	None						
	Time too Short	Case too shallow, soft Hub	6		Wrong Set up	1	Cause Control: Set up Instruction, Inspection Plan Characteristic Control: PLC Controls, Part Position, Power Supply Regulation, Quench Control,	1	6	None						
			6		Coil Deterioration	1	Cause Control: Standard Operator Instructions, Inspection Plans	1	6	None						

Fig. 14.3 Process FEMA. Source: Ref 2

Characteristic ID/Name	Potential Failure Mode	Potential Effect(s) of Failure	Sev	Class	Potential Cause(s) of Failure	Occ	Current Process Controls	Det	RPN	Recommended Action(s)	Responsibility & Target Completion Date	Action Results Actions Taken	Sev	Occ	Det	RPN
Quenching																
Concentration of Polymer/Water Mix	Too Lean	Quench Cracks	7		Measuring Error	1	Cause Control: Set up Instructions, Inspection Plan	1	7	None						
	Too Rich	Shallow Case Depth	6		Measuring Error	1	Cause Control: Set up Instructions, Inspection Plan	1	6	None						
	Flow Too Low	Shallow case depth, Soft spots	6		Improper Set up	1	Cause Control: Set up Instructions, Inspection Plan	1	6	None						
	Time too Short	Shallow case depth, Low Hardness	6		Improper Set up	1	Cause Control: Set up Instructions, Inspection Plan	1	6	None						
	Temperature too High	Shallow case depth, Low Hardness	6		Improper Set up	1	Cause Control: Set up Instructions, Inspection Plan	1	6	None						
Tempering																
(batch in furnace)	Not Tempered	Surface too hard	1		Load not put in furnace	1	Cause Control: Instructions, Inspection	1	1	None						
	Wrong Temperature	Wrong hardness	1		Controller Set Wrong	1	Cause Control: Instructions, Inspection	1	1	None						
	Time too short	Surface too hard	1		Controller Set Wrong	1	Cause Control: Instructions, Inspection	1	1	None						

Fig. 14.3 (continued) Process FEMA. Source: Ref 2

Table 14.1 Model of quality induction hardening system

Machine design	Process controls	FMEA, DCP, inspection	Preventive maintenance	Training

Specifications may include
 	
1. Define quality requirements
2. Select power supply
3. Define and select cooling water system
4. Define quench system requirements
5. Define heat station requirements
6. Define control system requirements
7. Specify system configuration
8. Specify any specific print and manual
9. Specify initial system startup needs and training
10. Warranty
11. Terms of purchase
	Minimum controls suggested
	
	1. Quench system
	  a. Pressure
	  b. Low flow
	  c. Low and high temperature
	  d. Filtering
	  e. Quench quality
	2. Cooling water system
	  a. Low pressure and flow: power supply, heat station, coil
	  b. High temperature
	  c. Resistivity
	  d. Filtering
	3. Control system for quality
	4. Induction coil
	  a. Ground fault detector if possible
	  b. Cooling water
	  c. Pressure switch for low pressure if pressure boosting pump is used
	  d. Workpiece position correct in coil
	5. Fixture
	  a. Rotation detection
	  b. Means of lubrication
	  c. Machine shutdown on fault and part rejection process
	  d. NDT quality verification if possible
	Verification
	1. Specified and docu-verification of trip conditions that cause a fault or trip
	2. Documentation and follow up on malfunction check sheets
	Frequency
	To be established after review of control plan
		1. Complete FMEA
		2. DCP, if required
		3. Define all inspection responsibility including short- and long-term capability studies
		4. Design inspection check sheet
		5. Set up production control process sheets
		6. Define any containment procedures, such as lock boxes for parts rejected by faults
			1. Create malfunction check sheets
			2. Verify check sheets are being used
			3. Create preventive maintenance program
			5. Quality standards
			6. System checks
				Who to set up what
				
				1. Introduction to metallurgy
				2. Process parameters
				3. Quality standards
				4. System checks
				
				Operators
				
				1. Process parameters
				2. Coil maintenance
				3. Quality standards
				4. System checks
				
				Maintenance
				
				1. Process parameters
				2. Coil maintenance
				3. System checks
				4. Troubleshooting

NDT, nondestructive testing. Source: Ref 2

of these risks. Figure 14.3 illustrates a FMEA for a specific induction-hardened part.

The goal is to produce defect-free, induction-hardened parts. Mistake proofing (Poka-Yoke) is a technique to eliminate errors often referred to as *failsafing*. The intention is that preventive techniques are used to control the repetitive production processing, with written forms used to present a plan that shows that the process has been designed properly.

A Quality Induction Heat Treating System

Quality assessment systems having previously been defined, the elements of a quality induction heat treating system certainly depend on the application and the system being used. Table 14.1 shows a model of the elements of a quality induction-hardening system that includes machine design, process controls, FMEA, and so forth, preventive maintenance, and training. This type of checklist helps in considering all aspects of the potential purchase of an induction heat treating system. Existing systems can also be compared so that possible system improvement can be made.

Process control involves monitoring all important variables that can affect the processing quality. Table 14.2 provides an example of a malfunction check sheet that can be used to record the processing variables with

Table 14.2 Malfunction check sheet

Induction Hardener No. _____   Date: _____

	Required setting	Actual setting	Fault indicator	Cycle stop	Remarks

1. Quench system
 a. Low-pressure fault at manifold
 b. Low quench flow
 c. Quench temperature low
 d. Quench temperature high
 e. Plugged quench holes on inspection
 f. Quench filter pressure low
 g. Quench concentration check
 h. Quench needs changing

2. Cooling water system
 a. Low-pressure/flow power supply
 b. Low-pressure/flow heat station
 c. Low-pressure/flow coil
 d. High-temperature power supply
 e. High-temperature heat station
 f. High-temperature coil
 g. Water resistivity low
 h. System filter needs change

3. Power supply energy output
SOLID STATE
 a. Record actual power output
 b. Record actual voltage output
 c. Record actual current output
 d. Record actual frequency
 e. Record any other system fault
 f. Record any system limit indication
RADIO FREQUENCY
 a. Record plate current
 b. Record plate voltage
 c. Record grid current
 d. Record RF overload fault
ALL
 a. Record energy monitor fault
 b. Record rotation slow fault
 c. Record any part proximity fault
 d. Record low temperature IR fault
 e. Record any other system fault

(continued)

Table 14.2 (continued)

	Required setting	Actual setting	Fault indicator	Cycle stop	Remarks
4. Induction coil					
a. Ground fault					
b. Coil overheating					
c. Coil buss, fishtails, quick connect overheating					
d. Coil alignment off					
e. Defective coil (visual) or inspection					
f. Coil flux concentrator defective					
5. Mechanical malfunctions					
a. Any fault					
b. Any breakdown					
c. Other					
6. Metallurgical Quality					
a. Surface hardness					
b. Case depth					
c. Pattern length					
d. Microstructure					
e. Other					

Source: Ref 2

their limits and settings. The fundamental concept of statistical process control (SPC) is that process adjustment is made on an ongoing basis to keep the process in control.

REFERENCES

1. Automotive Industry Action Group (AIAG), unpublished data
2. R.E. Haimbaugh, Induction Heat Treating Corp., personal research

CHAPTER **15**

Maintenance

MAINTENANCE and preventive maintenance programs are important factors in keeping induction heating systems in operation. When troubleshooting is required, trained personnel should be used because of the high voltages present inside the power supplies (particularly radio frequency, or RF power supplies). Most maintenance for induction heat treating systems is centered on the cooling water systems, the high-frequency electrical output connections from the heat station, the induction coil, any mechanical fixturing, and the quench system because the power supplies and heat stations have very few mechanical devices,

Maintenance requirements should be determined with a preventive maintenance program established according to the needs of the system. Table 14.1 as presented in the Chapter "Quality Control" in this book provides a model for maintenance requirements. Radio frequency power supplies, for instance, have hour meters so that oscillator tube life can be monitored for warranty purposes. The preventive maintenance (PM) schedule should list the items to be monitored and the frequency for monitoring. Critical items that require daily monitoring may need their own check sheets. Periodic items, such as those monitored monthly or semiannually, may need to have work orders generated so that the maintenance can be performed and verified. Individual manufacturers' maintenance programs should be the minimum level to which maintenance programs are followed. This Chapter will present recommendations for general maintenance.

Power Supplies and Heat Stations

Solid state and radio frequency (RF) power supplies need little periodic maintenance, assuming that they are in dust-tight enclosures. Experience has shown that a monthly visual inspection of the inside of the power supply is beneficial. The examination should include inspection for water leaks from hose connections, and signs of discoloration on copper buss, tubes connections in oscillator sets, and any signs of oil leaks from capacitors. The studs and taps on any heat station component, such as capacitor or transformer, need to have special attention paid to overheating, plus cleaning if dirty or corroded. Flow and pressure switches should be checked for proper operation on a semiannual basis. Separate heat stations should also be monitored.

Maintenance and Replacement of Capacitors

Any sign of an oil leak is considered an immediate cause for replacement of a capacitor. Shorted capacitors must be removed. At times faulty capacitors can be difficult to diagnose when the shorts are intermittent, with the capacitor "healing" itself between heat cycles. If one side of a capacitor is not grounded, then the capacitor must be discharged before handling. An insulated wire should be used to avoid drawing the discharge from the threads of the capacitor terminals, which might result in damage to the threads. When removing and installing capacitors, care must be taken so that the studs are not over-torqued. Disposal of any old capacitors should be reviewed to see if there are any federal, state, or local regulations that are applicable.

High-Frequency Electrical Output

This is considered to be anything in the high-frequency electrical connections from the terminal of the output transformer in the heat station to the induction coil. Coils have leads that are used to fasten onto either an output buss connection or the heat station transformer output. Because output buss and coil leads are exposed to the quenchant, there is a tendency for buildup of scale particles from the workpieces and subsequent arcing. Depending on the quenchant, periodic inspection and cleaning should be done. Gummy residue as left from polymer quenchants should be removed. All connections should be examined for signs of discoloration caused by overheating. When coils are changed, the electrical connections should cleaned. Mild abrasives such as steel wool or scotchbright (or equivalent) should be used to prevent stock removal. If there are any bolt replacements, the bolts must be made of the same nonferrous material as the bolts previously used; or if stainless, the stainless must be non-magnetic. When fastening the output buss or coil into the transformer, to keep from stripping the holes the bolts must be carefully torqued according to the manufacturer's instructions.

Induction Coil

Many induction coils are low maintenance, particularly the coils that are used on lower power densities and those without magnetic flux concentrators. If any leaks develop, the coils should be removed from production, cleaned, and repaired. Coils with magnetic flux concentrators need to have periodic inspection, sometimes daily, for degradation and deterioration of the concentrators. Some of the coils that operate with high-current power densities flex during operation, leading to fatigue such as failures or cracks. Coils usually fail at a stress point, where there is an abrupt change in direction of the coil, such as a right angle with a brazed joint. Also, some coils develop restrictions to the cooling water flow and start to vibrate due to steam pocket formation of the cooling water during the heat cycle. This is an indication that failure is imminent.

Cooling Water Systems

Each manufacturer has recommendations for water system purity for their power supplies, generally with conductivity, dissolved minerals, solids, and pH specified. The purity of the cooling water systems must be maintained. If the cooling systems are closed systems, such as deionized systems with automatic control of the water resistivity, a neutral pH of 7 together with recommended biocide levels should be maintained. General operating specifications are discussed in the Chapter "Induction Heat Treating Systems" in this book. A good control range for the water resistivity is 50/100 μS/cm, while some feel that a range of 10 to 400 μS is suitable. If there is not closed-loop control of water purity, the water should be periodically changed according to the instructions of the manufacturer, or every six months at a minimum. Filtering down to 25 μm is advised, and of course, means should be present for the cleaning and/or exchange of filters on a logged basis. Table 15.1 provides troubleshooting for closed water systems with towers.

Table 15.1 Water system troubleshooting

Fault	Possible problem	Preventive maintenance (PM)
Line water is too hot	Outside temperature is too hot for tower	Be able to add city water if necessary
	Tower fan has failed	
Power supply water too hot	Line water to heat exchanger is too hot	Monitor line water temperature.
	Heat exchanger is fouled	Clean heat exchanger on PM program
	Temperature controls not working	Repair. Have emergency bypass available
	Temperature control valve not working	Repair. Have emergency bypass available
Individual line overheating in power supply or heat station	Clogged water line	Clean. Check for proper filtering and resistivity
Low pressure out of filter	Clogged filter	Clean or replace
High return pressure (reduces flow through power supply)	Return water line clogged	Clean
	Return water line too small	Increase pipe size

Source: Ref 1

Quench System

Quench rings, pads, and integral quench coil outlets are prone to clogging when either unfiltered or polymer-based quenchants are used. Some of the different types of polymer quenchants tend to build up with use, coating not only anything in contact with the quenchant, but also tending to close and restrict the outlet holes. When a particular outlet hole clogs, the flow of quenchant will redistribute to the other outlet holes. The result can be incomplete quenching of the workpiece, although the total flow of quenchant is the same. The flow redistributes through the other holes, and flow switches may not indicate the clogged condition. Daily visual inspection is sometimes the only means of finding clogged holes, and inspection should be done with the quenchant flowing. If individual clogged holes cannot be cleaned, the quench ring may need to be removed for cleaning.

Maintenance requirements change according to the type of system and quenchant. Over a period of time all quenching systems need to have the heat exchangers cleaned. Unless otherwise established, this should be done every year. Oil quenches may have Y-type or screen filters that periodically need to be cleaned. In addition, a sample of the oil should be sent back to the supplier for testing at least once a year to make certain that there has been no deterioration. Water and polymer quench systems have five basic causes of contamination:

- Solids contamination. Scale and metal particles collect in the quench tank and should be removed on a continuous basis.
- Tramp fluid build-up can occur from extraneous sources such as machining fluids on the workpieces, hydraulic oil leaks from hydraulic lines in the fixturing, and excess grease/oil that was applied to conveyors, bushings, and other lubrication points. These contaminate the water-polymer and can change the cooling rate while the refractometer produces readings that appear to make the concentration of water-polymer in range. Viscosity is used as a control method to help determine accurate concentrations in these cases and is discussed later in this section.
- Biological contamination is likely in any polymer quenchant system. The contamination is most likely to be noticed after a period of inactivity, such as over a weekend. There will be an odor to the quenchant with possibly a "slime" covering the surface. Microbiological treatment can be effective, but should be under the guidance of an expert.
- Dissolved materials occur in some regions of the country in the make-up water used to mix polymers. Some evaporation of water occurs during quenching, and over a period of time the dissolved salt level increases. The nature and effect of dissolved salts will depend upon the make-up water used. Over a period of time the build-up of dissolved

salts can affect the cooling rate of the quenchant and, if not recognized, may not be identified by normal refractometer or viscosity tests. The faster cooling rate can produce undesired metallurgical effects for some steels and parts such as quench cracks or distortion.
- Breakdown of the polymers can occur over time of use. The polymers differ in their stability and bonding, and they are subject to thermal/oxidation conditions that can lead to polymer degradation. The effect is that although the concentration does not change, the quenching rate changes. This problem seems to be application sensitive, and many systems last indefinitely while other systems may suddenly start cracking the workpieces. The degradation cannot be determined by viscosity or through use of a refractometer. If quench cracking becomes a problem, recharge the quenchant to see if the cracking is eliminated. If so, the polymer quenchant should be periodically recharged. One simple way to do this is to determine either a time period or production rate of the number of parts needed to be run before recharging.

Rust on Workpieces

Closed loop, water based and polymer quench systems need to have the rust inhibitors checked on a weekly basis.

Periodic Maintenance

All filters must be cleaned or changed. In addition, the concentration of the water/polymer ratio needs to be verified at least weekly through a refractometer and/or viscosity. Figure 15.1 illustrates the viscosity/concentration relationship, with the multiple curves reflecting the flexibility in temperature at which the viscosity determination can be made. Because the viscosity of the quenchant solution is largely independent of other factors, such as solids contamination, a viscosity measurement provides a better reading of the effective polymer ratio. It should be noted that there is continued development for using the viscosity as an on-line means of concentration verification.

Overtemperature alarms or shutdown controls installed on quench systems should be tested on a six-month basis to make certain that protective devices are operating. In addition, quality control methods should be established to ensure that there is no contamination of the quenchant. Oil quenches can be contaminated by water with the bath itself appearing coffee colored rather than the dark, black, or bluish appearance of pure quench oil. Polymer-water quench systems that are contaminated by

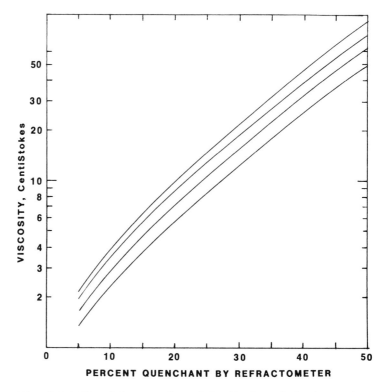

Fig. 15.1 Viscosity/concentration relationship for polyglycol quenchant. Source: Ref 2

hydraulic fluid tend to discolor the quenchant, and in the morning before the system is started up will be floating on the surface of the quench tank. Air can also be a contaminant in a water-base quenchant, giving it a foamy or milky appearance. Contaminated quenchants of any kind should be replaced with new quenchant.

Mechanical Systems or Fixturing

The mechanical systems or fixturing will have individual needs according to the particular system needs, including lubrication, adjustment, cleaning, and so forth. With the increased use of programmable controllers, relay operation is not a problem. However, there may the solenoids and other types of proximity switches and limit switches that need cleaning and maintenance.

REFERENCES

1. R.E. Haimbaugh, Induction Heat Treating Corp., personal research
2. R. Blackwood and E. Mueller, Installation and Control of Polymer Quenchants, *Heat Treat.*, Oct 1981

APPENDIX 1

Metallurgical Definitions for Induction Heat Treating

A

alternating current (ac). An electric current that reverses its direction many times every second. The current supplied by the utilities in the United States reverses 60 times per second and is called Hertz (Hz). Power supplies used for induction heat treating change the 60 Hz line power to much higher frequencies, most commonly ranging from 3 kHz (3,000 Hz) to 450 kHz (450,000 Hz).

anneal. To soften.

austenite. The microstructural phase that is formed when steel is heated above the transformation temperature (approximately 760 to 1010 °C, or 1400 to 1850 °F, depending on the type of steel). When steel is held above the transformation temperature long enough, the transformation to austenite will be completed. It is necessary to induction austenize steel before it can be quenched. The term *induction harden* really means to induction austenize and quench so that the area that is heated is hardened.

B

bainite. A mixture of iron carbide plates in ferrite that forms when steel is not quenched fast enough. Bainite usually forms during cooling below the 565 to 275 °C (1050 to 530 °F) temperature range. Bainite is not a desirable product in induction-hardened parts that specify the microstructure be 100% martensite.

boiling stage. The second stage of cooling when quenching a part. (The first stage is the vapor stage.) The quenchant that is in contact with the part boils until the temperature of the surface of the part passes below the boiling temperature of the quenchant being used.

Brinell. A hardness tester that uses balls to make an impression. A measurement of the diameter of the indent is made. A table is used to convert the diameter of the indent in millimeters to the Brinell hardness number (BHN). The test is often used by heat treaters to measure the hardness of annealed and normalized steels as produced by conventional heat treating furnaces. The Brinell numbers can be correlated to Rockwell numbers with BHN numbers of 230/400 correlating to HRC 20/43.

BHN. Brinell hardness number. *See Brinell.*

C

capacitor. An electrical device that stores a charge of electricity. Capacitors with taps are used to change the frequency for tuning power supplies.

carbide. A compound of carbon and a metal such as iron. Carbon in steel is normally found as iron carbide. The objective in heat treating is to dissolve the carbides during austenitization. Because of this it is recommended that the carbides be dispersed in the steel in small particle sizes.

carbonitriding. A process in which heat treating furnaces produce a thin layer of iron carbide and nitrides on the surface of a steel part. When quenched, the layer is very hard. The case depth may be 0.0762 to 0.3048 mm (0.003 to 0.012 in.) deep. Nitrides make the parts resistant to induction tempering. Therefore, parts that have been carbonitrided may be difficult to induction temper or anneal below 50 HRC.

carbon. Carbon is an element. This means that it cannot be broken down into any other form. It can exist in the three different microstructures—carbon, graphite, or diamond.

carbon restore. The process of restoring a decarburized surface to the core carbon content. This is a batch process and is done in an atmosphere heat treating furnace with special atmosphere control.

carbon steel. Steel that contains up to 2% carbon with no other alloying elements.

carburizing. A process in which a heat treating furnace is used to produce a layer of iron carbide on the surface of a steel part. When quenched the layer is very hard. The case depth may be from 0.127 to 2.54 mm (0.005 to 0.100 in.) deep. Steel parts can be first carburized and then either quenched and tempered or slow cooled before being induction hardened.

cast iron. Iron that contains carbon content higher than 2%. The excess carbon will be found in the form of iron carbide or graphite. The types of cast iron most likely to be processed with induction are gray cast iron, nodular cast iron, and modular cast iron.

cementite. See *iron carbide.*

Charpy impact test. An impact test in which a V-notched, keyhole-notched, or U-notched specimen, supported at both ends, is struck behind the notch by a striker mounted at the lower end of a bar that can swing as a pendulum.

cold drawn bars. Most bars less than 38.1 mm (1.5 in.) in diameter are cold drawn. The bars are first hot rolled and wound into coils. Then the bars are straightened and are pulled through dies that reduce the diameter of the bar.

compound. A compound is formed when two or more elements combine. An example is iron carbide. A common example of another compound is calcium carbonate (TUMS), which is taken for stomach acid.

compression. The force that occurs when an object is pushed together from both sides. Some induction-hardened parts form compressive cases that actually add to the strength of the part.

conductor. An object that transfers heat or electricity. Heat flows through, or is conducted by, metal. Electricity flows through a copper wire conductor. A part being heated by induction is conducting electricity poorly (there are high heat losses due to the resistance to the flow of the high frequency electric current).

continuous cooling diagram. The diagram that shows the microstructures that are produced by different rates on continuous cooling for a type of steel.

convection. The transfer of heat by conduction in a fluid such as air, water, or oil.

convection stage. The last stage of cooling when quenching, in which heat is transmitted from the surface of a part by convection through the quenchant. This is the slowest portion of quenching a part. This is also where most, if not all, of the transformation to martensite occurs.

core hardness. The hardness found near the center of a bar.

critical temperature range. The range in temperature at which austenite starts to form and completes forming.

cryogenic treatment. The cooling of steel parts in freezers ranging from -129 to -185 °C (-200 to -300 °F).

crystal. (Grain) Small units of atoms that are aligned alike. When viewed under the microscope the crystal looks the same. Crystal surfaces can be seen when parts are broken.

Curie temperature. The temperature 766 °C (1413 °F), at which steel becomes nonmagnetic. Radio frequency (RF) induction heaters will have the plate current drop. Solid-state induction heaters will have an automatic frequency shift to regulate for constant output.

current density. The amount of current in an area (amps/square inch). The higher the current density, the faster the rate of heating.

cut off wheel burn. The localized tempering effect that can be produced by a cut off wheel. This can be due to factors such as improper cooling, not positioning the part correctly, or using the wrong wheel material (i.e., too hard of a cut off wheel material for the hardness of the part being cut).

D

decarburization (decarb). A surface condition on steel parts for which the surface has lower carbon content. The carbon has burned out during the original making or subsequent heat treating of the steel. A heat treated part with this condition will have lower hardness on the surface. This condition is normally fixed by either restoring the carbon in a furnace or machining off the lower carbon surface.

ductility. The property of a metal that allows it to be permanently deformed before rupturing.

E

electromagnetic. The invisible energy field that surrounds an induction coil.

effective case depth. The depth of the hardened case is defined by a hardness value. HRC 50 is a commonly used value; however, other values, such as HR15N scales and microhardness scales, may be used. Because there is no agreed standard for induction-hardened case depths, many manufacturers define their own standard.

etch. Solutions of water or alcohol with a small amount of nitric or hydrochloric acids attach to the surface of metal to show grain boundaries and compounds such as iron carbide (pearlite).

eutectic. The chemical composition of 0.8% carbon where austenite, on slow cooling, forms ferrite and carbide simultaneously. Note that this is the lowest temperature for transformation to austenite on heating. The transformation occurs at one temperature, and there is no transformation range.

F

ferrite. Ferrite is thought of as pure iron, and it is one of the two crystalline forms assumed by iron.

ferrite banding. Parallel bands of ferrite that run in the long direction of a bar of steel. The ferrite bands are between bands of pearlite.

flashpoint. The point at which an oil quench will catch fire and stay on fire. Below this temperature a flame caused by a part being quenched will self extinguish.

flux field. Refers to the invisible lines of electromagnetic energy surrounding an induction coil.

frequency. Frequency means the same as Hertz. It is how many times an electric current reverses its direction in a second. Frequency was the term used in the United States before Hertz was adopted. Older literature will refer to 60 cycles rather than 60 Hertz or 450 kilocycles (Kc) rather than 450 kHz.

G

graphite. Graphite is one of the forms of carbon that is found in ductile cast irons. It is very soft. Cast irons with graphite will produce lower Rockwell test readings than the hardness of the iron itself because of the graphite's effect in lowering average hardness. Microhardness is used to measure the hardness of iron in the matrix.

grains. Metals are crystalline in nature. In order to see grains under a microscope, the parts are polished to eliminate scratches and then etched. Etching shows the grain boundaries.

grain size. The number of grains in a given area. Grain size numbers range from 1 to 15, with 1 being the largest grain size.

grinding crack or burn. The pressure of a grinding wheel on a hardened part can produce intense heat at the point of contact. This heat cannot only locally temper the part but can also produce small cracks. The cracks often form networks that look like chicken wire. Tempering parts above 121 °C (250 °F) before cutting or grinding reduces the tendency toward production of grinding cracks. Grinding or cut-off wheel burns can be detected by double etching-techniques as outlined in specimen preparation techniques.

H

hardenability. The term for how deep below the surface a part will harden. Hardenability is important for the selection of steels that are capable of quenching cases deep enough to meet specific case depth requirements.

hardenability bands. The minimum and maximum hardness that is expected from commercially available heats of steel from end-quench testing procedures. The end-quench is commonly called the *Jominy hardenability test*. The hardness is measured starting from the end of the bar at 1.59-mm (1/16-in.) intervals. The distances of interest for induction hardening is the first 1.59 mm (1/16 in.) because it represents the range of surface hardness that can be expected. The other distance of interest is the depth at which HRC 50 can be produced because it represents the effective case depth that can be expected.

heat of steel. Refers to steel that is made in one batch. The chemistry and microstructure should be the same.

heat station. The name given to the enclosure placed between induction heater (power supply) and the induction coil. The heat station contains any necessary transformers and capacitors for load matching and tuning. Some power supplies have the heat station built in, and others require external heat stations. The induction coil is attached at the output side of the heat station.

Hertz (Hz). Hertz means the same as *frequency*. The line frequency furnished in the United States is 60 cycles.

high frequency. When discussing power supply frequencies, frequencies greater than 100 kHz.

homogeneous. Uniform. A homogeneous microstructure is desired in parts for induction hardening.

hot rolled steel. Hot rolling first reduces all steel bars in cross section size.

HRC, HRA, HR15N, HR30N, HR45N, HRB, RC, RA, and so forth. Rockwell hardness test scale indicating which particular test scale is to be used.

I

induce. When one conductor of electricity causes an electric current to flow in another conductor. An induction coil induces current into the part being heated.

induction coil or inductor. Water-cooled copper windings that circulate the electric current for induction heating. See also *work coil*.

induction heater or power supply. The name given to the electrical box which changes the 60-Hz line frequency to higher frequency.

iron. Pure iron is one of the elements. See also *ferrite*.

iron carbide. A compound of iron and carbon that is very hard and brittle.

isothermal transformation diagram. This diagram does not give much information for the purposes of induction quenching. The diagram shows the microstructures produced by rapid quenching to specific temperatures, and then holding at that temperature until transformation from austenite is complete.

Izod impact test. A type of impact test in which a V-notched specimen, mounted vertically, is subjected to a sudden blow delivered by the weight at the end of a pendulum arm.

L

low frequency. Refers to frequencies between 1 kHz and 10 kHz.

M

macrostructure. The metal structure that is seen with no magnification. The part may be polished and etched. Sometimes flow lines from forging and some grain and phase structures can be seen without magnification.

magnaflux. An inspection process that uses ultraviolet light to detect cracks in a part.

manganese. An element that is present in concentrations from 0.6 to 1.5% in steels that are normally induction hardened. This element helps steel to harden better.

martensite. The microstructure that is desired when quenching a part. The normal objective is to produce a fully martensitic microstructure if possible.

martensite transformation range. Depends upon the type of steel. During quenching, transformation to martensite generally starts at or below 265 °C (530 °F) and completes at room temperature or slightly below.

medium frequency. Frequencies between 10 kHz and 100 kHz.

melting. Melting is caused when parts are overheated. After cooling the surface will appear to have flowed.

microhardness. A hardness indent is made with a microhardness testing machine and is measured at 400 times magnification.

microstructure. The crystalline structure and phases that are seen on a polished and etched part when viewed by a microscope. Common magnifications used are from 100 to 1000 times.

N

normalize. The process of heating a part over the critical temperature and cooling at a specified controlled rate.

O

oil quench. Special oils that are made to controlled chemistry and that have a high flashpoint (do not catch fire easily). There are several manufacturers, and the oils can be purchased to different quenching speeds.

output transformer. Common name used for the potted transformer to which the coil leads are attached on RF induction heaters. Most medium-frequency and low-frequency induction heaters used for heat treating also use output transformers for voltage change and tuning.

P

pearlite. Fine pearlitic microstructures are desirable for parts before being induction hardened. Pearlite is produced when steel that has first been austenitized is then cooled slowly through 760/482 °C (1400/900 °F). The austenitic microstructure transforms in iron carbide platelets (cementite) in a matrix of iron (ferrite). Pearlite is not normally found in the martensite in induction-hardened areas.

phase. Grain or crystal with the chemical composition and atomic alignment the same. Metallic crystals are solid and do not have the transparency of crystals such as diamonds.

phase diagram. The iron-carbon phase diagram is the diagram used to define what phases are present in steel for all carbon concentrations at different temperatures.

pitting. Overheating causes pitting. As steels are heated in air greater than 760 °C (1400 °F), the scale on the surface becomes thicker. As the temperature of the heated part increases, scale formation increases to the point that the scale will start flaking. With temperatures greater than 980 °C (1800 °F), the scale can be thick enough that on a finished part there is deterioration of the surface finish. Small pits or holes are formed that can be seen after the parts are quenched.

polymer quench. Mixture of a small percent of a polymer (when concentrated from its container it pours like glue) that is put into water to reduce the quenching speed. The higher the concentration or temperature, the slower the quenching speed. With higher concentrations there is more drag-out to the point that the surface of the part quenched will feel slimy. All polymer quench residue should be removed before further processing of a part.

power supply. See *induction heater*.

precipitate. When a substance that is dissolved in a liquid comes out of solution in the liquid. Polymer quenches have dissolved polymer. When parts are first quenched and in the vapor stage of quenching, a thin film of the polymer precipitates (forms) on the surface of the part. This film produces uniform, fast quenching. When the part cools so that the quenchant is in the boiling phase, the polymer redissolves into the quenchant.

Q

quench. To cool a part. Common quenchants are air, oil, water, and various polymers. There are three stages of quenching: vapor, boiling, and convection.

R

resistance. Refers to how much a conductor resists the flow of an electric current. All metals are conductors. Copper has low resistance and is a good conductor. Steel has much higher resistance and is a poor conductor.

retained austenite. Not generally desirable in parts that are induction hardened. Caused by overheating some steels such as American Iron and Steel Institute (AISI) 52100 and some of the tool steels. All of the austenite does not turn into martensite during quenching. Retained austenite has a hardness of about 40 HRC. Retained austenite can be transformed to martensite through combinations of tempering and cryogenic treatment.

resulfurized steel. Steels such as AISI 1141 and 1144 contain a high sulphur content purposely added during the making of the steel. Sulphur combines with manganese to create a soft compound that makes the steel more machinable.

Rockwell. A hardness testing machine that indicates hardness by the depth to which a small diamond penetrater will penetrate with a given load.

S

scale. Steel parts form a black iron oxide when heated in air. Above approximately 818 °C (1500 °F), scale starts to flake. When quenched the scale looks like a black, flaky material on the part.

scan. See also *progressive harden*. The power is left on the power supply while parts are moved past the coil with continuous application of quench.

seam. A shallow crack that is formed when the steel bars are made. Seams will tend to be 0.05 to 7.62 mm (.002 to 0.300 in.) depth. They run along the length of a bar. Seams may fade in and fade out. There may be more than one seam. Seams are considered material defects. Bar stock that has seams must have the seams machined or ground off. Seams can open up more during quenching and give the initial impression that the part has a quench crack.

secondary quench. Refers to a second quench that is applied at a fixed time after a part is taken out of the primary, or first quench. In scanning operations a second quench ring may be used.

segregation. Non-uniform distribution of an element, phase, or impurity. Ferrite banding is an example of segregation.

single shot. See also *static*. Refers to the act of heating the entire part on a timed basis. The induction coil may enclose the entire part or a section of the part. If possible the parts are usually rotated when enclosed, and must be rotated or moved when partially enclosed.

spheroidical. Round, such as small round carbides (sometimes referred to as globular) that are desirable in the microstructure for AISI 52100 steel prior to induction hardening.

spotty hardness. Islands of low hardness, which can occur when quenching with water. Steam pockets can occur during quenching in small spots. This causes the steel to cool more slowly in these regions, producing incomplete transformation to martensite.

stainless steel. Steel that has chromium added to increase the resistance to rusting. The types of stainless steel are ferritic, austenitic, and martensitic. Martensitic stainless is the only stainless that is magnetic and is the series of stainless steels that are induction hardened (such as 416, 430, and 440 stainless steels).

static heat cycle. Same as single shot. Refers to the act of heating the entire part on a timed basis. The induction coil may enclose the entire part or a section of the part. If possible, the parts are usually rotated when enclosed, and they must be rotated or moved when partially enclosed.

stress. A surface or internal force of compression or tension in a part. A part can be in tension at one place and in compression in another.

stress relieve. Heat in a draw furnace at temperatures between 121 to 593 °C (250 to 1100 °F) to reduce or homogenize internal stresses.

T

temper. Heat in a draw furnace at a temperature between 121 to 593 °C (250 to 1100 °F) to reduce the hardness of a quenched part.

tempered martensite. Martensite that is tempered from anywhere between 121 and 593 °C (250 and 1100 °F). The higher the tempering temperature, the softer and more ductile the part.

tension. The act of pulling apart. Parts crack when a tensile force exceeds the tensile strength of a part.

tool steel. Elements such as chromium, nickel, molybdenum, titanium, tungsten, and vanadium are added to the steel to make the steel harden better and to give special wear and strength properties.

torsion. The act of twisting a part.

total case depth. The case depth as defined by etching. Glass bead or nital etching are the two most common methods used.

transformation product. Usually refers to products other than martensite that are formed when a part is not quenched fast enough to produce a completely martensitic microstructure.

transformation range. The range of temperature within which austenite forms either on heating or on cooling. This range is different (higher) for heating than the range for cooling. On phase diagrams the starting point for transformation on heating is referred to as the Ar_1 and the point for completion is the Ar_2. On cooling the starting point is referred to as the Ac_1 and the point for completion is referred to as the Ac_2. Note that these points are defined by cooling very slowly and establishing the points at which the transformations occur.

transformation temperature. The temperature at which a phase change occurs. The temperature at which ferrite starts to transform to austenite is commonly referred to as the *transformation temperature*. Note that at the eutectic point of 0.8% the transformation starts and completes. At other chemical compositions there is a transformation range.

transformer. An electrical device that has an input current and voltage which are changed into a different current and voltage. (If the voltage rises, the current decreases and visa versa.) Transformers with taps are used for tuning.

transition product. Undesirable products that form when parts are not quenched fast enough. Examples are pearlite and bainite.

V

vapor phase. The first phase of quenching a hot part. The quenchant next to the part forms a vapor.

Vickers. The type of microhardness indent that is diamond shaped.

W

work coil. Water-cooled copper windings that carry the electric current that causes a part to heat. Same as *induction coil*.

APPENDIX 2

Scan Hardening

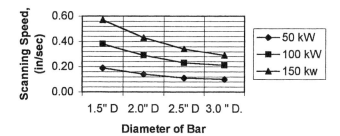

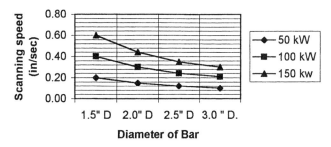

Fig. A2.1 Scan hardening for 3 kHz. Source: Ref 1

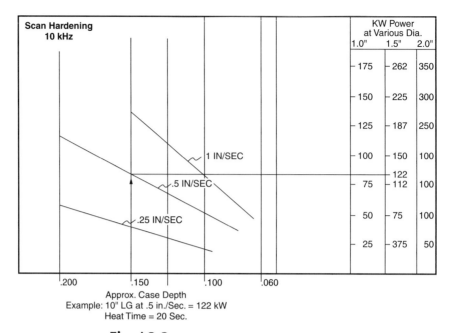

Fig. A2.2 Scan hardening 10 kHz. Source: Ref 2

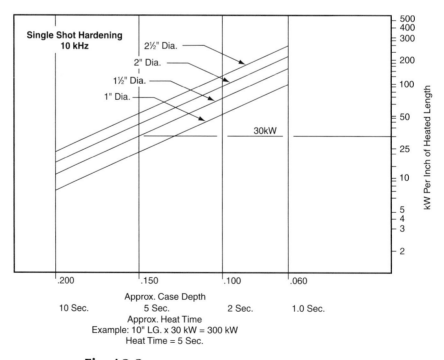

Fig. A2.3 Single-shot hardening for 10 kHz. Source: Ref 2

Section size		Material	Frequency(a), Hz	Power(b), kW	Total heating time, s	Scan time		Work temperature				Production rate		Inductor input(c)	
								Entering coil		Leaving coil					
cm	in.		Hz	kW	time, s	s/cm	s/in.	°C	°F	°C	°F	kg/h	lb/h	kW/cm²	kW/in.²
Rounds															
1.27	½	4130	180	20	38	0.39	1	75	165	510	950	92	202	0.067	0.43
			9600	21	17	0.39	1	510	950	925	1700	92	202	0.122	0.79
1.91	¾	1035 mod	180	28.5	68.4	0.71	1.8	75	165	620	1150	113	250	0.062	0.40
			9600	20.6	28.8	0.71	1.8	620	1150	955	1750	113	250	0.085	0.55
2.54	1	1041	180	33	98.8	1.02	2.6	70	160	620	1150	141	311	0.054	0.35
			9600	19.5	44.2	1.02	2.6	620	1150	955	1750	141	311	0.057	0.37
2.86	1⅛	1041	180	36	114	1.18	3.0	75	165	620	1150	153	338	0.053	0.34
			9600	19.1	51	1.18	3.0	620	1150	955	1750	153	338	0.050	0.32
4.92	1¹⁵⁄₁₆	14B35H	180	35	260	2.76	7.0	75	165	635	1175	195	429	0.029	0.19
			9600	32	119	2.76	7.0	635	1175	955	1750	195	429	0.048	0.31
Flats															
1.59	⅝	1038	3000	300	11.3	0.59	1.5	20	70	870	1600	1449	3194	0.361	2.33
1.91	¾	1038	3000	332	15	0.79	2.0	20	70	870	1600	1576	3474	0.319	2.06
2.22	⅞	1043	3000	336	28.5	1.50	3.8	20	70	870	1600	1609	3548	0.206	1.33
2.54	1	1036	3000	304	26.3	1.38	3.5	20	70	870	1600	1595	3517	0.225	1.45
2.86	1⅛	1036	3000	344	36.0	1.89	4.8	20	70	870	1600	1678	3701	0.208	1.34
Irregular shapes															
1.75 to 3.33	¹¹⁄₁₆ to 1¹⁵⁄₁₆	1037 mod	3000	580	254	0.94	2.4	20	70	885	1625	2211	4875	0.040	0.26

(a) Note use of dual frequencies for round sections. (b) Power transmitted by the inductor at the operating frequency indicated. This power is approximately 25% less than the power input to the machine, because of losses within the machine. (c) At the operating frequency of the inductor.

Fig. A2.4 Typical operating conditions for progressive through hardening of steel parts by induction. Source: Ref 3

REFERENCES

1. R.E. Haimbaugh, Induction Heat Treating Corp., personal research
2. Tocco Inc., 30100 Stephenson Hwy, Madison Heights, MI 48071-1677
3. *Heat Treating*, Vol 4, *ASM Handbook*, ASM International, 1991

APPENDIX 3

Induction Coil Design and Fabrication

THE CHAPTER "Induction Coils" in this book shows examples of the types of induction coils that are typically used. While some of the simple coils are easy to make and use, other coils such as machined coils need to have development programs with testing before the final coil design is produced. This section will discuss materials used in coil building, coil design, and coil construction techniques. In order to help understand coil design, the electromagnetic flux fields and their influence on coil design will be presented. In addition, magnetic flux concentrators and their uses with induction coils and computer-aided design for induction coils also will be covered.

Induction Coils

Coil Materials. Coils are made from copper because of its low resistance to the flow of the high frequency current and its relatively low price. Coils made from tubing and many of the low- and medium-power coils are made from commercial copper, while oxygen-free high-conductivity (OFHC) copper is favored for the high-power, machined coils. When high power is sent through these coils, it is important that coil losses are minimized. All induction coils usually require water cooling to absorb the power losses and to prevent overheating of the coils.

One of the design limitations of coils is the inability of the water passages to carry enough water for proper cooling. Tube inside diameter (ID) limits minimum copper tubing size that can be used. Tubing is made as small as 2.6 mm (3/32 in.) in diameter, while the maximum size is limited more by the ability to shape and form the tubing rather than by size. Tubing sizes commonly used are shown in Table A3.1 with other sizes available. Tubing should be purchased in the annealed state if possible.

Table A3.1 Copper tubing commonly used for induction coils

Tubing Type			
Round		Square/Rectangular	
mm	Inches	mm	Inches
3.125	$\frac{1}{8}$	3.125 × 3.125	$\frac{1}{8} \times \frac{1}{8}$
4.688	$\frac{3}{16}$	4.688 × 4.688	$\frac{3}{16} \times \frac{3}{16}$
6.250	$\frac{1}{4}$	6.250 × 6.250	$\frac{1}{4} \times \frac{1}{4}$
9.375	$\frac{3}{8}$	6.250 × 12.50	$\frac{1}{4} \times \frac{1}{2}$
		12.50 × 25.00	$\frac{1}{2} \times 1$
		12.50 × 37.50	$\frac{1}{2} \times 1\frac{1}{2}$

Source: Ref 1

Materials for Coil Support and Fixtures. The materials used around coils should be either nonconductive or nonferrous. Small brass screws, with their heads removed, can be brazed onto the copper turns with the threaded ends inserted into fiberglass supports secured with nuts. Aluminum channels are easy to use for support guides. Trays and wheels for holding workpieces often use Transite 2, a non-asbestos insulating material. While steel is used for holding nests if sufficiently far from the flux field, brass and nonmagnetic stainless steel is often used when the flux fields heat the nests. Some setups even provide a quench outlet for fixture or nest cooling during the heat cycle. Table A3.2 shows a number of the materials that have been used over the years.

Table A3.2 Structural materials for induction heating fixtures

Material	Characteristics and comments
Nonmetals	
Diamonite(a)	Aluminum oxide; very hard, dimensionally stable; standard and special shapes; resists high temperatures
Epoxy Fibre Glass FF91(b)	Coil-mounting supports; good electrically; also useful for soft soldering to about 230 °C (450 °F)
Transite II(c)	Work-table tops; heat resistant; avoid for coil supports (electrically poor)
Mycalex, Supramica(d)	Various grades; useful for high-temperature coil supports; special shapes available
Silicone rubber (RTV)(e)	Molds accurately; flexible; useful for soft soldering to 315 °C (600 °F)
Fired Lava(f)	Easily machinable before firing; good heat resistance; less strength than Diamonite
Nonmagnetic Metals(g)	
Aluminum alloys	Fixture base plates, work tables; useful for soft soldering (does not bond)
Brass (free machining)	Supporting screws, locators adjacent to coils, sinks, etc; corrosion-resistant pins; quench and recirculating tanks
Titanium	Useful for positioning of parts to be silver brazed; does not stick
Nichrome	Excellent high-temperature strength; oxidation resistant; used for locating and holding of parts
Inconel alloys	Useful for locating parts, radiant (susceptor) heating of thin materials
Magnetic Metals(g)	
Low-carbon steel	Structural members; cabinets; work tables
Alloy steel (hardenable)	Moving parts of fixtures subject to wear
Stainless steel (hardenable)	Moving parts of fixtures subject to wear and corrosion

(a) Diamonite Products Div., U.S. Ceramic Tile Co. (b) Formica Div., American Cyanamid Co. (c) John-Manville Co. (d) Mycalex Corp. of America. (e) Dow-Corning Co. (f) American Lava Corp. (g) Materials listed are available from a number of manufacturers. Source: Ref 2

Water Flow in Coils. Coils with inadequate water flow will melt or will have soldered or brazed joints fail. Care must be taken in the fabrication of the coils to minimize restrictions. As with all piping used to carry water, the water passages should be clean, free of burrs, and able to avoid reduction in cross section, such as at brazed joints. The cooling water, if free of dissolved solids, can have the temperature rise to over 80 °C (175 °F). Therefore, compared to the cooling water requirements of the power supply that may permit only an 80 °C (20 °F) temperature rise, the amount of water in gpm is reduced. The amount of cooling water needed for a coil is linear to the temperature rise and can be calculated: kW = gpm × Δ°F × 500 × 3414. If the coil loss is 50% of a 100 kW system and a temperature rise of 45 °C (85 °F) is permitted, the coil water flow required is approximately 4 gal/min. When the coil cooling water, even if aided by a booster pump, cannot provide this flow, the coil will overheat and fail. Steam pockets, which occur when water is heated to the melting point, produce coil overheating and eventually a melting failure. Figure A3.1 shows water flow versus tubing inside diameter (ID). Tube cross-sectional area, length, and restrictions such as right-angle turns affect the amount of flow through a given coil. One technique that is used to increase the water flow to higher than that provided by round tubing is to use rectangular or square tubing. Rectangular tubing is more difficult to wind around small diameters without collapsing, and to wind multiturn coils with the turns characterized for either different diameter spacing or offset between the turns. Techniques such as pouring sand or even pouring a low melting point eutectic alloy in the coils have been used to help prevent collapse.

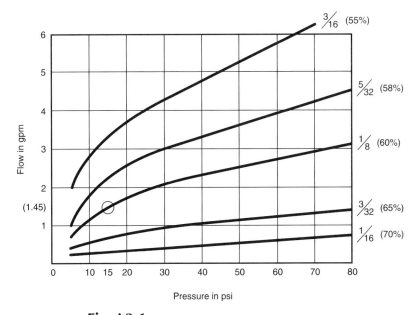

Fig. A3.1 Water flow through tubing. Source: Ref 3

(The eutectic has a melting point below 100 °C, or 212 °F and can be blown out when submerged in boiling water.)

Coil Wall Thickness Adjacent to the Workpiece. The current in a coil is concentrated mainly on the face of the coil that is adjacent to the workpiece. Therefore, the amount of copper in this part of the coil will dictate how much the copper heats due to the resistance of the high-frequency current. Reference depth and the depth of heating are discussed in the Chapter "Theory of Heating by Induction" in this book. Because the reference depth increases as frequency decreases, the most efficient ideal coil will have wall thickness as shown in Table A3.3 for guidelines.

All of this having been said about tubing cross section, much of the time the coil design prevents a wall thickness that is less than ideal from being used. Thinner wall thickness is sometimes needed on high-power density coils because the copper itself cannot conduct heat fast enough to the cooling water to prevent overheating. When this occurs, the wall thickness must be reduced.

Coil Efficiency Design. The coupling or air gap between the coil and the workpiece has a large overall effect on coil efficiency. With higher frequencies, the gap is more important because the coil's magnetic flux field is coupled more tightly to the coil. Therefore, the closer any coil is coupled to the workpiece, the higher the coil efficiency. Thus while tight gaps are important, in practice the process itself determines the coupling as a result of the spacing needed to keep the workpiece from hitting the coil. Coupling distances can vary from as close as .75 mm (0.030 in.) to 25 mm (1 in.). Loose coupling during austenitizing is used only when necessary for the process. An example is a workpiece that is scanned and has two different diameters to be heated. The coil is coupled as tightly as possible to the largest diameter. The coupling then increases when scanned onto the smaller diameter. Another requirement for using larger coil coupling is the warping of scanned parts during heating and the subsequent need for larger coil coupling to keep the coil from arcing to the part. Finally, when the workpiece being heated has areas such as fillets that need to be hardened, the coil coupling may be changed to promote uniform heating of the shaft diameter and the fillet area. Typically the coupling needs to be close to the fillet and adjacent shoulder.

Table A3.3 Copper coil wall thickness selection for frequency

	Minimum Wall Thickness	
Frequency	mm	inches
450 kHz	0.8 to 1.0	0.032 to 0.040
10 to 50 kHz	1.0 to 2.1	0.048 to 0.090
3 kHz to 10 kHz	3.0 to 4.0	0.125 to 0.156

Source: Ref 1

Coil Design. The statement used to be made that coil design is 50% theory and 50% practice. Experienced fabricators of induction coils can make most of the common coils without any difficulty. If a new application is involved, design of the induction coil starts with the analysis of the workpiece and production requirements, as discussed in the Chapter "Induction Heat Treating Process Analysis" in this book, with a suitable frequency and output power selected.

After frequency selection and power, quenching and fixturing are built around the coil design. Sometimes more than one coil design can produce the desired result. An example of this occurs when an encircling coil is used to scan a part or where a longitudinal, single-shot coil is used to statically heat the same workpiece. When high production rates are needed, the choice may be multiple heating positions or the single-shot heating. Single-shot heating has limits in not being able to heat long parts because the energy requirements are too high. Long parts, such as axle shafts, are scanned. As mentioned mechanical handling equipment can require a larger air gap that can influence frequency and even power choices. The high frequencies, such as 450 kHz, are most efficient with tight coupling to the coil. The lower frequencies are more tolerant to wider coupling to the coils. Table A3.4 shows recommended coil coupling for efficiency.

Coil Turns. The relationship of coil type and coil efficiency is important. For instance, where optimal efficiency is desired, a two-turn helical coil is preferred over the use of a single-turn helical coil. Multiturn coils are even more efficient, and helical coils are more efficient than the hairpin or channels coils. Finally, internal coils tend to be the most inefficient, particularly on small bores. This makes bores less than 12.5 mm (0.5 in.) hard to heat by radio frequency (RF), and bores less than 25 mm (1 in.) hard to heat with 10 kHz.

Multiturn coils are the most efficient. The number of turns on a coil is dictated by the ability of the heat station to either match the coil voltage directly or to have an output transformer that will match the turns. The higher voltages and low current of RF permits more frequent use of multiturn coils. In the same respect, 3 kHz through 25 kHz coils are more likely to be low-turn, heavier-current carrying coils. The single-turn type coils can be machined to make them more rigid. Machined coils also have the ability to use integral quenches, and this can make them useful in some systems.

Table A3.4 Coil coupling to workpiece versus frequency

Frequency	Coil Coupling to Workpiece	
	mm	inches
50 to 450 kHz	1.5 to 2.0	0.060 to 0.080
3 to 50 kHz	2.0 to 3.0	0.080 to 0.120
1 kHz to 3 kHz	3.0 to 6.0	0.120 to 0.230

Source: Ref 1

Radio frequency single-turn encircling coils have an electromagnetic dead spot at the rear of the coil where the coil leads attach to the coil. Workpieces must be rotated to provide uniform heating because of this dead spot. As the frequency of the power supply lowers, this dead spot becomes less pronounced. At 10 kHz encircling-type coils built to the contour of the cam will heat the cam without a dead spot.

Different workpiece shapes also dictate frequency selection. For instance, rounded-square shaped power transmission shafts will uniformly heat with a helical encircling coil on 10 kHz, while the same coil design will heat only the corners on 450 kHz. At 10 kHz, the coil can be a helical coil with the shaft rotating, rather than contoured to a square shape.

Tube or Machining. The first basic decision in material selection for a coil is whether to use tubing or to machine the coil. Tubing is generally much more economical to use, and it is often used for higher-frequency and multiturn coils. Round, square, and rectangular tubing is used. Copper tubing is wound and is contoured to the desired shape. When winding, any time work hardening becomes excessive the tubing must be annealed by heating it red-hot and quenching in water. Copper is "hot short," so it will break when winding red-hot unless special forming techniques are used. Where coil efficiency is important, square and rectangular tubing are more efficient. However, they are more difficult to form without collapsing. Copper tubing is silver brazed to the coil leads. If the leads cannot provide sufficient coil rigidity, brass studs can be brazed onto the coil turns and fastened into insulated blocks. Liners can be brazed into copper tubing when heating the workpiece. These liners serve to make the coil act as if it were a square or rectangular coil. The liners also absorb mild arcs, preventing water leaks that would have originated from a tubing coil. An example is a 1.5- by 122.5-mm (1/16 by 0.5 in.) copper insert that is brazed into the 75-mm (3-in.) ID of a 6.25-mm (0.25 in.) round copper encircling coil. The coil induces current into the workpiece as if the coil was 12.7 mm (0.5 in.) wide.

Machined coils make use of copper buss bar or plate that has cooling passages drilled or milled. The machined cooling passages then have precisely made copper inserts brazed over the passages. Machined coils can use both machined sections and tube sections brazed together

Connection to Leads. All coils need to have leads that connect to the output power section of the heat stations. Where the distance to the coil is small, the coil either bolts directly onto the terminal output of the heat station, or slides into a quick connect mechanism that is already attached. When the distance is extended, buss transmission leads are generally needed. All fasteners used must be nonmagnetic. Brass, silicon bronze, and, where higher strength is necessary, nonmagnetic stainless steel bolts are used for the connection of the coil or buss leads. Hose clamps, if used, need nonmagnetic screws. There are a number of different types of bolt and hole configurations, depending upon the power supply and manufacturer. Before ordering any coil, the coil manufacturer must know what design is being used.

Bus and Lead Design. The design of the coil leads and extension bus is important for most power supplies in order to deliver full power to the coil, except for some power supplies that need additional resistance in the coil circuit in order to make full power. As the output power and distance increase, the bus leads are made wider with a narrow gap in between. Many bus leads have a 1.5 mm (0.060 in.) air gap with a Teflon insert between the leads. Illustrations of some bus extensions are shown in the Chapter "Induction Coils" in this book under the coil discussion.

Coil Characterization. When designing a coil, it is important to visualize the distribution of the eddy currents that cause the workpiece to heat. The objective is to induce current in the area required to be heated. The specifications of the workpiece will show the desired heat treated pattern. In effect, the coil needs to be designed to produce the high heat in the pattern area. Single-turn coils produce eddy-current distribution dependent upon the coupling to the workpiece. Figure A3.2 shows the temperature profile resulting from the electromagnetic flux in a simple, single-turn coil. Current concentrations are less intense at the ends and provide taper in the profile. The right-hand figure shows the profile from a conditioned coil that was machined to produce offset in the center of the coil. By changing the coil coupling, the temperature profile is less intense in the center and more intense on the edge, producing a more uniform and wider profile. Another way to characterize a coil is to add turns. Figure A3.3 illustrates four examples of coil characterization on a multiturn coil: Figure A3.3(a) shows the uneven heating produced in the original coil. Figure A3.3(b) spreads the middle turns, decreasing the coil efficiency in the center. Figure A3.3(c) changes the coil coupling by making the center turns tighter and the outside turns looser, and Figure A3.3(d) simply makes the

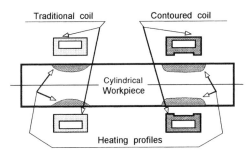

Fig. A3.2 Induction heat treatment: temperature profiles with traditional and contoured single-turn induction coils. Source: Ref 4

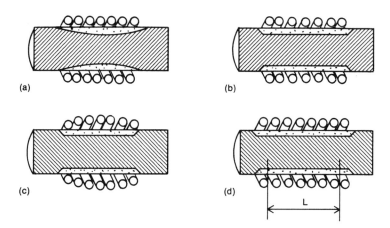

Fig. A3.3 Characterization of multi-turn coils. (a) uneven heating pattern in a round bar obtained by a coil with an even pitch, a problem which can be corrected by: (b) increasing the pitch of the central turns of the coil, (c) varying the coupling, or (d) using a longer coil. L, the length to be heated. Source: Ref 5

coil longer by adding a turn. The other type of coil modification is to add a flux concentrator to the coil. This will be discussed later.

Quenchant Effect on Coil Design. The type of quenchant and how it is to be applied needs to be known before coil design can be completed. As discussed in the Chapter "Quenching" in this book in the selection of quenchants, there are three basic types of quenching:

- Integral, from within the coil
- Spray, separate from the coil and
- Submerged or dunk, separate from the coil.

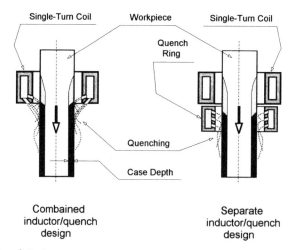

Fig. A3.4 Combined and separate coil/quench design. Source: Ref 4

Table A3.5 Workpiece diameter versus quench-hole size

Workpiece Diameter		Quench Hole Size	
mm	inches	mm	inches
6.25 to 12.5	0.25 to 0.50	1.15 to 1.58	0.046 to 0.063
12.5 to 37.5	0.50 to 1.50	1.58 to 2.35	0.063 to 0.094
greater than 37.5	greater than 1.50	3.13 to 3.9	0.125 to 0.156

Source: Ref 1

The type of quenchant is dictated by the workpiece and process requirements. Once the quenchant has been selected, the method of application can be decided. The only real effect on the coil occurs when a spray quenchant is to be applied from the coil itself. Figure A3.4 shows a coil with integral quench and a coil with a separate quench. In this case, the coil is most likely to be a machined coil, because either the quenchant itself can be used to cool the coil, or a two-chamber coil can be made with one cooling passage for the coil and one for the quenchant.

The design of the quench inlets and outlets are discussed in the Appendix "Quench System Design" in this book. Integral quench coils have the quench holes drilled through the heating face. These quench holes should not take up more than 10% of the heating face. In order to control the direction of the quenchant stream, the wall thickness of the copper must be at least 1.5 times the orifice diameter. Table A3.5 shows rule of thumb quenchant hole sizes for different shaft diameters.

Design for Use of Flux Concentrators

Flux concentrators are used to shape the flux field so that more intense flux concentrations are produced in a workpiece in the areas where heat is required, while reducing flux leakage into areas in which heat is not desired. Flux concentrators improve overall process efficiency, but there are some heat losses produced in the concentrator material that must be removed. The coil cooling water or the quenchant must remove this heat. In addition, the electromagnetic forces of the concentrators produce forces that tend to loosen the concentrator from the coil. Due to the copper work hardening because of mechanical stresses caused by the coil flexing, if the original coil is susceptible to stress cracking, the use of any of the non-laminate concentrators enhances early coil failure.

Installation depends upon the type of concentrating material that is used. Laminates of various shapes and thicknesses can be purchased to various sizes and shapes (Ref 6). Laminates must be securely fastened to the coil in such a manner that there is no electrical path between the laminates. One method of installation is to wedge packets of laminates between copper wedges that are brazed to the coil. The concentrating materials, such as Ferrotron or Fluxtrol, can be purchased and can be screwed or epoxy glued to the coils after being machined to the correct contour (Ref 7). The moldable concentrator material can be purchased from Alpha 1 and is molded and baked to the coil.

Cooling must occur from conduction of the heat into the work coil, the quenchant from drilled holes through the concentrators where the concentrator is used for quench outlets, or from application of quenchant at the end of the heating cycle. Some of the concentrating materials have better resistance to deterioration from the quenchant if they are painted with glyptol.

Selection of material depends upon frequency, power density, and application. The concentrator suppliers have their concentrating materials rated by the intended frequency of use. Laminations are best used under 10 kHz, where they tend to be the most efficient. However, laminates do not work well around bends in the coil because of the problem of being able to stack them tightly. Also, laminates do not wear well in areas where they are the subject of flux fields that are produced where the coil has two different orientations, such as with some of the single-shot coils, which heat both longitudinally and around diameters at the ends of the coils. The machinable concentrators are suited for the types of coils in which they can be machined for good contact and for use with medium frequencies. The efficiency of these concentrators increases with frequencies from 10 to 50 kHz. The ferrite concentrators work best with the RFs. Care needs to be taken when using concentrating material in the RF regions where the dielectric properties of the concentrating materials are not high enough to prevent arcing. The moldable concentrating materials work well with many coils in which it is more difficult to machine the material. As of the date this book is written the claims regarding operating efficiency in general application of the different types of materials, such as the machinable versus moldable, are hard to resolve. Furthermore the expectancy is that continued improvements will be made in material for higher frequency use in the future.

Design. The concentrators are used to shape the flux field into specific areas. Figure A3.5 shows the shaping effect for a single-turn coil. The pattern produced by the coil with concentrating material is more uniform and shaped to the width of the coil than the pattern produced by the bare coil. Figure A3.6 illustrates the shaping of the flux field on a two-turn coil. The two-turn coil with concentrator produces more heat in the center area desired, with less stray flux. Figure A3.7 shows the flux field in an application such as a crankshaft journal. Use of concentrating material forces all

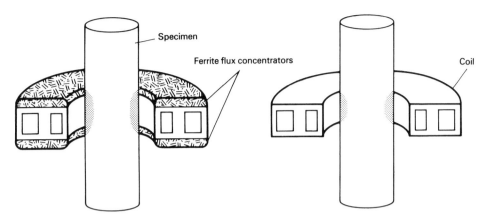

Fig. A3.5 The effect of using a flux concentrator on the heating of a shaft. Source: Ref 8

coil current into the bottom of the coil, inducing the current into the main diameter without heating the side.

Levitation of Workpiece. The magnetic flux field produced by high power densities can levitate the workpiece during heating. The levitation effect is more pronounced with the frequencies below 10 kHz, but it has been observed even at radio frequencies. Depending on the application, the workpieces may need to be held on both the top and bottom, or somehow otherwise clamped.

Computer-Aided Coil Design (CAD)

Computer simulation of induction hardening requires special software and considerations to achieve accurate results. When processing induction hardening involves several different phenomena that depend upon one another. With today's computer software and hardware, it is impossible to simulate all of these phenomena simultaneously. Hence, decisions must be made regarding what part of the problem is most important for simulation and what part must be done empirically or based upon experience. The best approach at the time this book is written involves electromagnetic and thermal field simulation, and if desired the thermal field can be exported to a structural transformation program.

The programs available for induction hardening simulation include:

- One-dimensional (1D) electromagnetic and thermal (coupled),
- Two-dimensional (2D) electromagnetic or thermal, 2D coupled,
- Three-dimensional (3D) electromagnetic or thermal, and
- For some specific cases, 3D coupled.

Due to the complexity of the electromagnetic field problem, no commercial computer simulation program exists for 3D coupled simulation of all

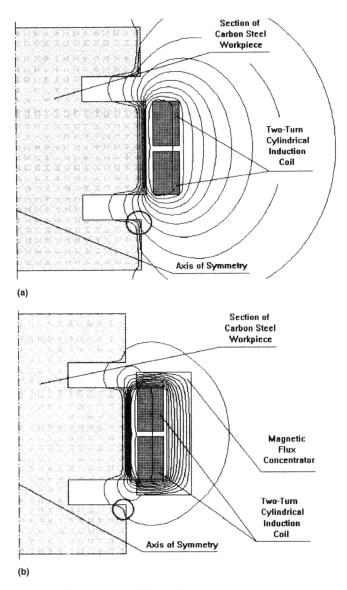

Fig. A3.6 Electromagnetic field distribution in two-turn coil (a) without magnetic flux concentrator, and (b) with magnetic flux concentrator. Source: Ref 4

induction system configurations. Also, the time that is required for direct 3D optimization of an induction hardening system would take from weeks to months, even with today's fastest computers.

Because of this, a hierarchical approach is currently taken to induction-hardening process design through computer simulation. The "rule of pyramid" for induction heating computer simulation states that more simple

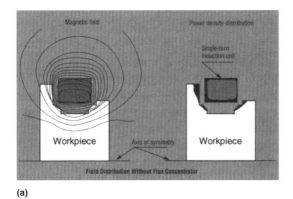 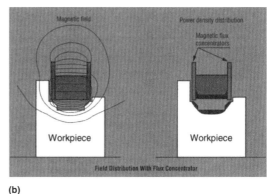

Fig. A3.7 Field distribution (a) without flux concentrator, and (b) with flux concentrator. Source: Ref 9

programs should be used, limiting the number of cases to be simulated with the more complicated software. It is believed that this approach, which begins with 1D coupled simulation, greatly reduces the time required for simulation of the induction-hardening process. For most cases, 1D coupled simulation alone or 1D coupled and some 2D simulation is enough for induction coil optimization.

ELTA was used to made a good approximation with the 2D system. Figure A3.8(a) and Fig. A3.8(b) show the temperature distribution in an induction scanning application. Figure A3.(a) represents the temperature distribution, and the lightest shade corresponds to the full austenization temperature required for complete martensite formation. In this process, induction was used to harden a 45.8 mm (1.8 in.) outside diameter (OD) shaft for the automotive industry. The frequency for this application was 3 kHz, and the scanning speed was 7.5 mm (0.3 in.) per second.

The results of induction hardening simulation can be almost exact. The accuracy of simulation depends mainly on the correct material property description, especially for magnetic steels. The software selection is critical also. Currently, there is a great deal of information available on the mechanical properties of steels at room temperature. However, there is a lack of information on material properties at elevated temperatures critical for induction heating simulation, such as permeability, electrical resistivity, and thermal conductivity. It is in the intermediate temperature ranges where the material database must be improved. Fortunately, above the Curie temperature, most steel alloys have close to the same electrical resistivity and thermal conductivity. Given accurate material properties, computer simulation can provide almost exact results for prediction and optimization of the induction hardening processes.

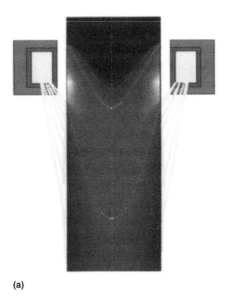

Fig. A3.8 ELTA, (a) simulation of induction scanning using ELTA, and (b) temperature graphs at different radii versus time for the induction scanning process using ELTA. Source: Ref 10

REFERENCES

1. R.E. Haimbaugh, Induction Heat Treating Corp., personal research
2. S.L. Semiatin and S. Zinn, *Induction Heat Treating,* ASM International, 1988
3. M. Black, "Advanced Coil and Tooling Design," paper presented at 6th International Induction Heating Seminar by Inductoheat (Nashville, TN), Book 2, Sept 1995
4. V. Rudnev et al., *Steel Heat Treatment Handbook,* Marcel Dekker, Inc., 1997
5. S.L. Semiatin and D.E. Stutz, *Induction Heat Treating of Steel,* American Society for Metals, 1986
6. Alpha 1 Induction Service Center, Inc., unpublished data
7. Fluxtrol, 1388 Atlantic Ave., Auburn Hill, MI 48326
8. *Heat Treating,* Vol 4, *ASM Handbook,* ASM International, 1991
9. V. Rudnev and R. Cook, Magnetic Flux Concentrators: Myths, Realities and Profits, *Met. Heat Treat.,* March/April 1995
10. V.S. Nemkov and R.C. Goldstein, "Computer Simulation for the Design of Induction Heat Treating Processes and Work Coils," paper presented at ASM Heat Treating Society Conference (Cincinnati, OH), ASM International, Nov 1-4, 1999

APPENDIX 4

Quench System Design

THE CHAPTER "QUENCHING" in this book covered the basics of quenching theory, including the selection of quenchants, their temperature control, and application. This section will discuss the design of the actual quench devices or outlets and the design of cooling systems. Quench outlets include spray-type quench fixtures, including various types of integral quenches in coils, quench rings, quench jets, and quench pads. The term *quench ring* will be used for describing the quenching fixture for the balance of this section. The quenching system contains the necessary controls, heat exchanger, temperature regulators, filters, and pumps necessary to provide cooling for the quenchant.

The Quench Cooling System

The design of the quench cooling system sets the basic limit on how much steel can be quenched. Other factors that influence the flow of quenchant on the workpiece are the hose inside diameter (ID) from the manifold to the quench ring, the number of hoses, the quench ring inlet size, and number of holes and hole size at the outlet of the quench ring.

The amount of quenchant required can be reasonably calculated from the theoretical viewpoint of assuming that the quenchant needs to extract all of the heat content needed to heat the workpiece. Other heat losses, such as residual heat left in the workpiece, are not significant. For example, if a workpiece requires an effective heat input of 40 kW, then the heat content for the quenchant can be derived by the method used to calculate the heat loss: gallons per minute (gpm) = $(40 \times 3413)/(500 \times 30)$ or 9.2 gal/min (from gal/min = $(kW \times 3414)/(500 \times$ temperature rise °F). For this 40 kW heat absorption the quench system must be capable of supplying 9.2 gal/min through the quench ring assuming a 22 °C (40 °F) temperature rise. Cutting the temperature rise in half would double the water flow requirement.

Table A4.1 Number of outlets to a quench manifold.

Quench manifold diameter		No. of manifold connections				
mm	in.	1	2	3	4	5
6.25	1/4	1/4	3/8	1/2	3/4	3/4
9.38	3/8	3/8	3/4	3/4	1	1 1/4
12.5	1/2	1/2	3/4	1	1	1
18.75	3/4	3/4	1 1/4	1 1/4	1 1/2	2
25	1	1	1 1/4	2	2	2 1/2

Source: Ref 1

Most quench systems use centrifugal pumps. Centrifugal pumps are sized by pressure drop and fluid flow requirement. The sizing of the pump versus the operating pressure is important not only for maximizing flow, but also for producing long motor life. A given centrifugal pump supplying quenchant into a manifold at a given pressure determines the maximum amount of quenchant flow that can be produced. With a given outlet pressure, the amount of quenchant that can be sprayed on a workpiece for cooling is dependent first on the design from the manifold outlets to the quench ring outlet holes, then on the quench ring design. As with any hydraulic system, system pressure drops reduce quenchant flow. Restrictions, sharp turns, corners, and smaller ID piping all reduce pressure and flow. In order to maximize flow, ideal changes in ID of piping outlets and piping size or hose should not be more than one to one. Thus, starting with the quench manifold outlets, the area of the total ID of the outlets from the manifold should not exceed the area of the ID of the manifold itself. If the outlet from the quench manifold (inlet to the quench ring or integral quench coil) has four 12.7 mm (0.5 in.) hose connections, the manifold needs to have a minimum of 25 mm (1 in.) ID. Table A4.1 shows the number of inlets that can be put into a quench ring from a given size manifold. There must be enough inlet lines of adequate size to the quench ring to furnish the needed quenchant flow. Figure A4.1 shows the quench flow

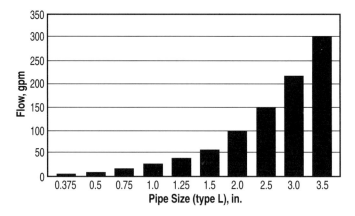

Fig. A4.1 Quench flow versus pipe size (velocity, 3 m/s, or 10 ft/s). Source: Ref 2

versus pipe size. Four 12.7 mm (0.5 in.) pipes would have a flow capability of 40 gal/min, while four 19 mm (0.75 in.) pipes would carry about 72 gal/min as shown.

The ability of a quenchant to absorb heat from a workpiece depends on the heat absorption characteristics of the quenchant itself, the effective flow of quenchant that is applied to a workpiece, the quenchant temperature, and the nature of the impingement (such as spacing and size of jets). The heat must be absorbed quickly so that spotty hardness and distortion do not occur. If sufficient quenchant is furnished from the manifold to the quench ring, then the size and number of quench holes at the outlet, along with the effective pressure at the coil, determine the amount of quenchant that can be applied. In order to assume that a workpiece will quench, a general recommendation is that the quench hole area should be a minimum of 5 to 10% of the area being quenched. As case depth and part diameter increase, the area of coverage needs to be increased. Rule of thumb for the basic minimum area is shown in Fig. A4.3, which indicates the amount of flow that can be applied depending on hole size and pressure. The rate of increase is roughly proportional to the pressure, in that doubling the pressure doubles the flow, while doubling the hole size such as from 1.59 to 3.18 mm ($1/16$ to $1/8$ in.) quadruples the flow. As mentioned under "Coil Design," the number of quench holes in an integral quench inductor is limited to 10 to 15% of the coil width so that there is sufficient copper in the coil to carry the current. The quench angle impingement commonly used is at 30° (60° impingement on the part). When separate quench rings and pads are used, this angle is sometimes changed to a straight 90° impingement. Table A4.2 shows an increase in quench hole size versus diameter.

From the determination of the amount of quenchant that can be produced at the inlet, the number of outlet holes can be calculated from the hole area and number of holes. Ideally the ratio of the inlet area to the total area of the outlet holes should be one to one. In practice, the ratio should never exceed one to two. Figure A4.2 shows the effect of inlet-to-outlet ratio on quenchant flow versus pressure. The hole-to-hole spacing is around 2 diameters, but the rows of holes may be staggered with offsets so that this can be increased. When the number of holes has been calculated, the design for drilling the holes can be planned. The previous example of 40 kW of heat put into the part required 9.2 gal/min of quenchant. Assuming that 5 mm ($1/8$ in.) holes are desired and spaced at 5 mm (0.25 in.) apart

Table A4.2 Workpiece diameter versus quench hole size

Workpiece diameter		Quench hole size	
mm	in.	mm	in.
2	$1/2$	$1 1/2$	$1/16$
4	1	3	$1/8$

Source: Ref. 1

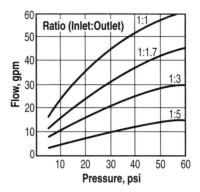

Fig. A4.2 Effect of inlet; outlet ratio on quench flow versus pressure. Source: Ref 2

in a 40 mm (2 in.) ID quench ring, there is a total of 25 holes per row for a flow at 20 psi (Fig. A4.3) of 44 gal/min. This is obviously more flow than is needed from a heat extraction viewpoint. The design might call for four to six rows of holes to first provide the impingement and coverage needed, and second to keep the part in the quench long enough. Obviously this raises the flow requirements greatly. However, sometimes the lower rows of holes can be spaced wider such as at 10 mm (0.50 in.).

The impingement of quenchant is most important in the stage of quenching that involves cooling below the "knee" of the time-temperature-transformation (TTT) diagram. Actual quenchant impingement on the workpiece is a function of the distance of the quench ring to the workpiece versus the spacing in the quench ring. With tight coupling, integral quench coils have impingement on the workpiece close to the spacing of the drilled holes in the coil. However, a separate quench ring, with loose coupling to the workpiece, will have the impingement spray converge. In this case, the drilled hole spacing around the ID can be increased. A second factor that seriously impacts the number of rows of holes needed is

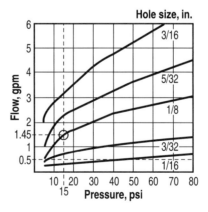

Fig. A4.3 Effect of hole size on quench flow versus pressure. Source: Ref 2

the time the workpiece needs to remain in the quenchant. The workpiece needs to be in the quenchant long enough for the entire case or area to transform to martensite without any tempering or draw back from residual heat in the core. As a rule of thumb, the minimum time for quenching on static heats is the same as the time required for heating. Parts that are being scanned need immediate quenching for two to three times the scanning time. In some cases additional quenching needs to be done through use of a secondary quench application below the first. With secondary quenches, often fixed lines or nozzles can be used because the workpiece is in or below the boiling stage of quenching.

Materials for Quench Rings. Any quench ring that is separate from the coil will heat if metallic and within the area of the induction coil's magnetic flux field. Therefore, the use of nonmagnetic materials is required. Surprisingly, if the quench ring spacing is kept distant from the coil, such as 25 to 50 mm (1 to 2 in.), and the quench is kept on any time there is a hot workpiece, plastics, fiberglass, and other synthetic materials work. The designs can be made either to fasten together so that any nonferrous screws are outside of the flux field, or so that the components can be glued together. Quench pads or blocks are made from copper or brass and placed outside the coil where the flux field intensity is low. Hose clamps (including the screw) and quick connects should be made from nonmagnetic materials if they heat from the coil's magnetic flux field.

Quench Cooling System Design

The quenchant must be selected for the characteristics desired. This includes the quenchant speed, desired operating temperature range, and additives. Some biocide additives contain chlorine which can combine with alkaline glycol to form cholorethel. End users can also specify nitrate- or non-nitrate based polymer quenchants to avoid the formation of nitrosamine (if nitrosamine is prohibited). Venting of the quench tank may be needed in cases where there is considerable vaporization and where the smell of the quenchant becomes obnoxious.

The quench cooling system must be sized correctly. A source of line water of the required flow and maximum inlet temperature is needed for cooling in the quench system heat exchanger. The quenchant must have the heat capacity to absorb the heat that is removed from the workpiece. Sizing of the system is a function of the fluid flows on both sides of the heat exchangers, temperature differentials of both fluids, and the size and heat-removal capacity of the heat exchanger. Cooling towers, chillers, and city or well water systems can be used for the line water. Cooling towers are discussed in the Chapter "Induction Heat Treating Systems" in this book and are most commonly used in many places when substantial line cooling water is needed due to cost. Cooling towers have the limitation of not being able to cool the recirculated water enough with high outside temperatures and high humidity. As with the quenchant system, closed systems involving chillers or cooling towers may need to have biocides added, and they may also need filtration to remove debris.

Plate-style heat exchangers, although more expensive than cartridge, are recommended. They are easy to clean, and extra plates can be added if additional capacity is needed.

Temperature-controlled valves are used to regulate the cooling of the quenchant by the line water in the heat exchanger. Inverse temperature control valves that operate proportionally by allowing only as much water as needed to provide very close temperature control can be used. Other systems use valves that open and close, as needed, to control the quenchant with a temperature range.

Quench systems must be able to provide the amount of quenchant needed. The system design must be capable of maintaining the quenchant within the specified control range. This range should normally not exceed more than 7 °C (10 °F) for a system. A normal control range might be from 35 to 38 °C (90 to 100 °F) or 38 to 443 °C (100 to 110 °F). When operating in controlled temperature ranges above room temperature, heaters are recommended to bring the tank up to temperature with an overtemperature alarm or cutoff used to prevent high-temperature operation.

The quench tank and piping must not react with the quenchant. Polymers are very hard on painted tanks. To prevent galvanic corrosion the piping should not be made of dissimilar materials. Tank storage capacity of three to four times the quenchant flow per minute is recommended.

The quench system piping, solenoids, and manifold must be designed so that the full volume of quenchant is supplied to the quenching position on command. Care must be taken so that the quenchant itself does not drain out of the quench ring. Full flow onto the workpiece should occur immediately after the quench solenoid opens, with no necessity to fill the quench ring first.

Suitable filtration must be put in the quench tank. The quenchant must be kept free of foreign material and scale. Y-Type, basket strainers are widely used and are the lowest cost. They trap only the larger particles and are suitable for quench tanks. In applications that use spray quenching, a higher degree of filtration is recommended to help keep the holes in the quench rings from clogging. In this case cartridge and bag-type filtration units can be specified to filter from 5 up to 100 μm. A differential pressure switch across the inlet and outlet sides will detect a clogged condition. If the filters are placed in a by-pass configuration (similar to the oil filter in a car), a separate flow is run through the cartridge continuously. When the filters are placed in series with the quench ring, a flow switch in series on the outlet side of the cartridge will indicate low flow. A wide variety of controls and systems are available to help "foolproof" these systems.

Continuous automatic indexing media style filters work well for users who have problems with manual replacement of filter media. These are sometimes called drag-out style filters because the filter paper is dragged across a perforated plate. When the pressure differential is too high, the filter is automatically indexed into a hopper. These filters are good for applications that are high in scale and chips because their large volume and automatic indexing assure good filtration. Positive pressure is used for filtration, and a clean tank is required on applications that cannot afford to have the quenchant flow disrupted in order to clean the filter.

Some of the power supplies, such as radio frequency (RF) oscillator supplies, have higher voltage at the terminal connection to the coil. Almost invisible layers of very fine magnetic scale from the quenchant can build up on these terminals, leading to arcing. Magnets or magnetic separators will remove these particles.

REFERENCES

1. R.E. Haimbaugh, Induction Heat Treating Corp., personal research
2. D. Williams, Quench Systems for Induction Hardening, *Met. Heat Treat.*, July/Aug 1995, p 34

APPENDIX 5

Induction Tempering

Metallurgy of Short-Time Tempering

Cellular manufacturing layouts require all processing for the manufacturing of a part to be in-line. When induction hardening is installed in-line and tempering is required, continuous tempering ovens were previously used. These ovens required substantial floor space. The use of induction tempering reduces floor-space requirements substantially. The metallurgical theory behind induction tempering is that short-time, higher-temperature tempering cycles can produce the same hardness as conventional long-time tempering. The higher tempering temperatures increase the diffusion-controlled precipitation process of carbon from martensitic microstructure. Figure A5.1 illustrates the effects of time and temperature for a 1050 carbon steel austenitized and quenched to a martensitic microstructure with a hardness of 62 HRC. A comparison of the hardness decrease is made between a one-hour furnace cycle and workpieces induction tempered for 5 and 60 s. For instance, 40 HRC is produced by furnace tempering at 425 °C (800 °F) or with a 5-second induction cycle heating to 540 °C (1000 °F).

Hollomon-Jaffe Tempering Correlation

Hollomon and Jaffe (Ref 2) investigated correlations for the time-temperature equivalence of different tempering processes. According to their formulation, the tempered hardness of martensite is a unique function of the quantity $T(C + \log_{10} t)$, where T is the absolute temperature, C is a material constant, and t is time in seconds. In the example shown by Figure A5.1 the constant C can be found by equating $(425 + 273)(C + \log_{10} 3600)$ to $(540 + 273)(C + \log_{10} 5)$, resulting in a value of C of approximately 17. With this value for C, equivalent time-temperature combinations for other hardnesses of the quenched 1050 steel can be obtained. For

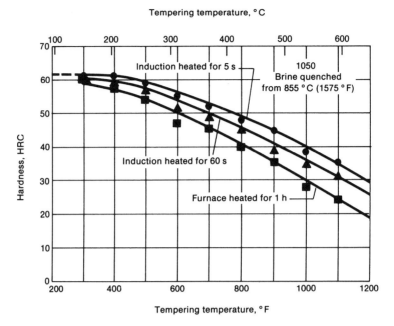

Fig. A5.1 The effects of time and temperature for a 1050 carbon steel. Hardness as a function of tempering temperature and time for furnace and induction treated 1050 steel austenitized at 860 °C (1575 °F) and quenched in brine. Source: Ref 1

instance, the 30 HRC tempered hardness obtained in a 1 h furnace treatment at 540 °C (1000 °F) may be obtained by a 60 s induction cycle of $(540 + 273)(17 + \log_{10} 3600)/(17 + \log_{10} 60) = 890$ °K = 625 °C = 1140 °F. This agrees with the data shown in Fig. A6.1.

The values of C in the Hollomon-Jaffe correlation for several plain carbon steels are given in Table A5.1. From this table it can be seen that the magnitude of C tends to increase with decreasing carbon content. In general, C tends to lie between 14 and 18 for steels with carbon contents below 0.5% and between 10 and 14 for steels with higher carbon content. Note that these values apply only when the tempering time, t, is expressed in seconds. If t is given in hours, the values of the constant need to be increased by 3.6 (approximately the $\log_{10} 3600$).

The values of C for plain carbon steels in Table A5.1 were taken from the work done by Hollomon and Jaffe. As determined by other research, the value of C can give predicted tempered hardness differences as much as one point HRC. This is within acceptable experimental accuracy and greater than the repeatability of hardness testing machines.

Extension of Tempering Correlation to Continuous Heating/Cooling Cycles. The Hollomon-Jaffe equation, although quite useful in conjunction with conventional tempering curves, should be applied with care for induction tempering. First, according to metallurgical theory, the

Table A5.1 Carbon parameters for several plain carbon steels

Steel	Starting microstructure	C
1030	Martensite	15.9
1055	Martensite	14.3
1074	Martensite	13.4
1090	Martensite	12.2
1095	Martensite	9.7
1095	Martensite + retained austenite	14.7
1095	Bainite	14.3
1095	Pearlite	14.1

Source: Ref 1

tempering temperature cannot be raised above the Ac_1 critical temperature, or transformation to austenite will start. Secondly, it must be realized that the correlation applies only to short-time tempering at a fixed temperature, rather than the heating to temperature cycles used by induction.

A means by which a particular time-temperature history is accounted for in the continuous rapid heating of induction can be derived by a simple extension of the Hollomon-Jaffe concept. This is done by calculating the equivalent time t^* for a constant temperature heating cycle which corresponds to the continuous cycle. One way of doing this is illustrated in Fig. A5.2. Here, the induction tempering cycle (shown schematically in Fig. A5.2a) consists of a heating portion and a subsequent cooling portion, the latter occurring at a somewhat lower rate. The total continuous cycle is broken into a number of very small time increments, each of a duration of Δt_I and characterized by some average temperature T_I. It is assumed that the temperature for the equivalent isothermal treatment is the peak

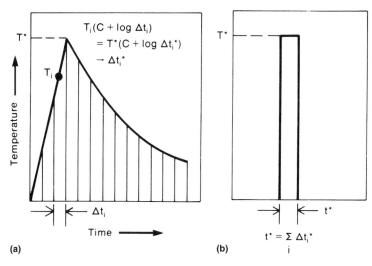

Fig. A5.2 Tempering and time broken into increments. Equivalence of (a) a continuous heating/cooling tempering cycles and (b) an "isothermal" treatment through the use of an effective tempering time (t^*) and temperature (T^*). Source: Ref 1

temperature of the continuous cycle, or T^*. However, this specification of the temperature for the isothermal cycle is arbitrary.

Having specified the temperature of the equivalent isothermal cycle as T^*, an effective tempering time t^* for this cycle can be estimated. This is accomplished by solving the increment in t^*, or Δt^*_I for each Δt_I in the continuous treatment by using the equation $T_I(C + \log_{10} \Delta t_I) = T^*(C + \log_{10} t^*_I)$. Summing the Δ^*_I for each portion of the continuous cycle yields the effective tempering time t^* at temperature T^* and, hence, the effective tempering parameter.

Semiatin and Stutz (Ref. 1), provide detailed information including a discussion of a correlation of fast tempering cycles by Grange and Baughman. The important correlation is that both Hollomon-Jaffe and Grange-Baughman suggest that the exact value of C has little effect and that the most important variables are the absolute temperature and the logarithm of the tempering time. Grange and Baughman also presented a means by which the tempered hardness in an alloy steel (10xx to 92xx series) could be estimated.

The methods for predication for an induction tempering cycle provide a good estimate for establishing the temperature parameters for the actual test temperatures of particular workpieces. The workpiece shape, direction of heat input, and area to be tempered will determine the actual temperatures required. Finally, as previously discussed, the desired final hardness specification must take into consideration that the Rockwell tester has a testing accuracy of plus or minus one point Rockwell C in the hardness ranges generally required from tempering.

REFERENCES

1. S.L. Semiatin and D.E. Stutz, *Induction Heat Treating of Steel,* American Society for Metals, 1986

APPENDIX 6

Tempering Curves

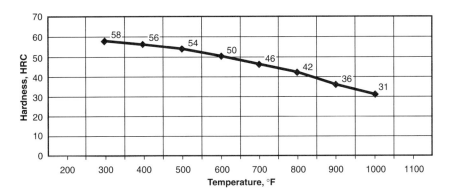

Fig. A6.1 Tempering curve for 1045 steel. Tempered 1 h at heat as quenched 60 HRC at 120 °C (250 °F). Source: Ref 1

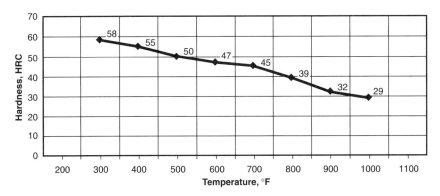

Fig. A6.2 Tempering curve for 1144 steel. Tempered 1 h at heat as quenched 60 HRC at 120 °C (250 °F). Source: Ref 1

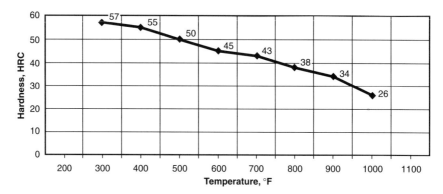

Fig. A6.3 Tempering curve for 4140 steel. Tempered 1 h at heat as quenched 60 HRC at 120 °C (250 °F). Source: Ref 1

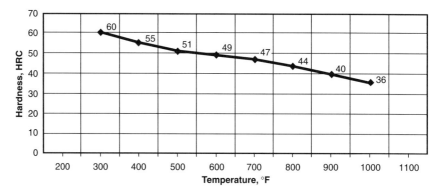

Fig. A6.4 Tempering curve for 4150 steel. Tempered 1 h at heat as quenched 62 HRC at 120 °C (250 °F). Source: Ref 1

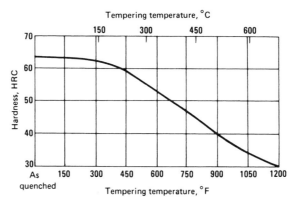

Fig. A6.5 Tempering curve for E52100 steel. Represents an average based on a fully quenched structure. Source: Ref 2

REFERENCES

1. R.E. Haimbaugh, Induction Heat Treating Corp., personal research
2. Standard Practices and Procedures for Steel, *Heat Treaters Guide*, American Society for Metals, 1982

APPENDIX 7

Hardenability Curves

SAE/AISI 1045H, UNS H10450

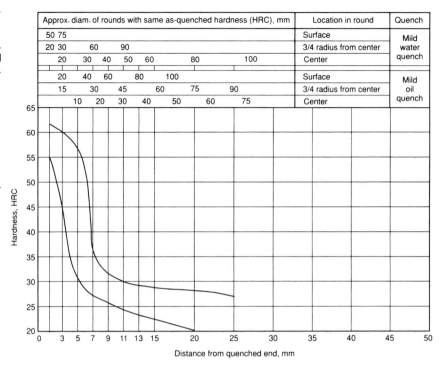

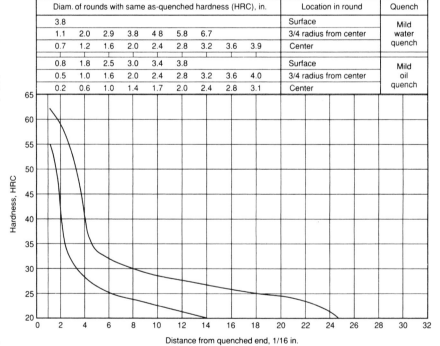

Fig. A7.1 Heat treating temperatures recommended by SAE. Normalize (for forged or rolled specimens only): 870 °C (1600 °F). Austenitize: 845 °C (1550 °F). Source: Ref 1

SAE/AISI 4130H, UNS H41300

Hardness limits for specification purposes

J distance, mm	Hardness, HRC Maximum	Hardness, HRC Minimum
1.5	56	49
3	55	46
5	53	40
7	51	36
9	48	32
11	44	28
13	41	26
15	39	25
20	34	24
25	33	23
30	33	22
35	32	20
40	31	...
45	31	...
50	30	...

Approx. diam. of rounds with same as-quenched hardness (HRC), mm	Location in round	Quench
50 75	Surface	Mild water quench
20 30 60 90	3/4 radius from center	
20 30 40 50 60 80 100	Center	
20 40 60 80 100	Surface	Mild oil quench
15 30 45 60 75 90	3/4 radius from center	
10 20 30 40 50 60 75	Center	

Hardness limits for specification purposes

J distance, 1/16 in.	Hardness, HRC Maximum	Hardness, HRC Minimum
1	56	49
2	55	46
3	53	42
4	51	38
5	49	34
6	47	31
7	44	29
8	42	27
9	40	26
10	38	26
11	36	25
12	35	25
13	34	24
14	34	24
15	33	23
16	33	23
18	32	22
20	32	21
22	32	20
24	31	...
26	31	...
28	30	...
30	30	...
32	29	...

Diam. of rounds with same as-quenched hardness (HRC), in.	Location in round	Quench
2 4	Surface	Mild water quench
1 2 3 4	3/4 radius from center	
0.5 1 1.5 2 2.5 3 3.5 4	Center	
1 2 3 4	Surface	Mild oil quench
0.5 1 1.5 2 2.5 3 3.5 4	3/4 radius from center	
0.5 1 1.5 2 2.5 3 3.5	Center	

Fig. A7.2 Heat treating temperatures recommended by SAE. Normalize (for forged or rolled specimens only): 900 °C (1650 °F). Austenitize: 870 °C (1600 °F). Source: Ref 1

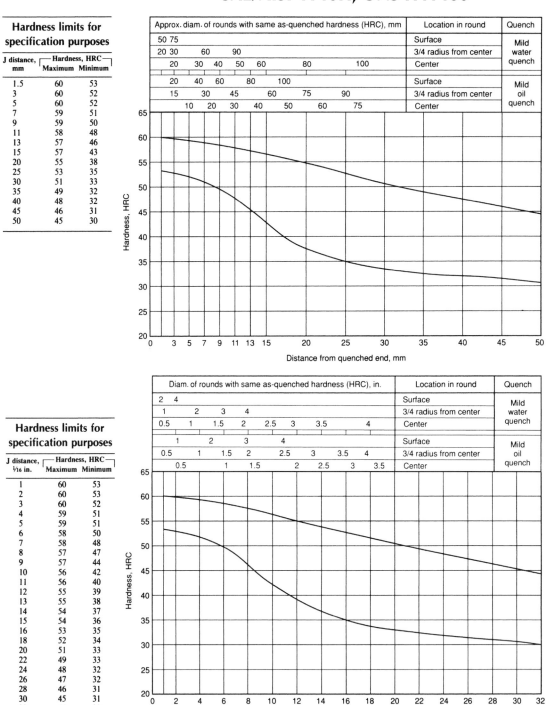

Fig. A7.3 Heat treating temperatures recommended by SAE. Normalize (for forged or rolled specimens only): 870 °C (1600 °F) Austenitize: 845 °C (1550 °F). Source: Ref 1

SAE/AISI 4142H, UNS H41420

Hardness limits for specification purposes

J distance, mm	Hardness, HRC Maximum	Hardness, HRC Minimum
1.5	62	55
3	62	54
5	62	54
7	62	53
9	61	52
11	61	51
13	60	49
15	60	48
20	58	43
25	56	39
30	55	36
35	53	35
40	52	34
45	51	33
50	50	33

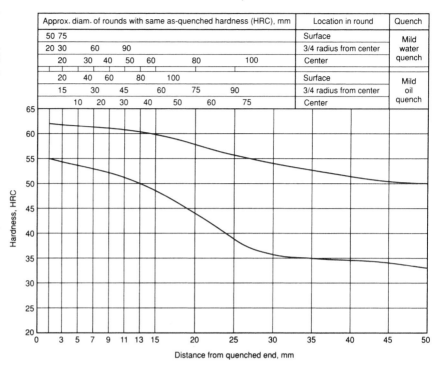

Hardness limits for specification purposes

J distance, 1/16 in.	Hardness, HRC Maximum	Hardness, HRC Minimum
1	62	55
2	62	55
3	62	54
4	61	53
5	61	53
6	61	52
7	60	51
8	60	50
9	60	49
10	59	47
11	59	46
12	58	44
13	58	42
14	57	41
15	57	40
16	56	39
18	55	37
20	54	36
22	53	35
24	53	34
26	52	34
28	51	34
30	51	33
32	50	33

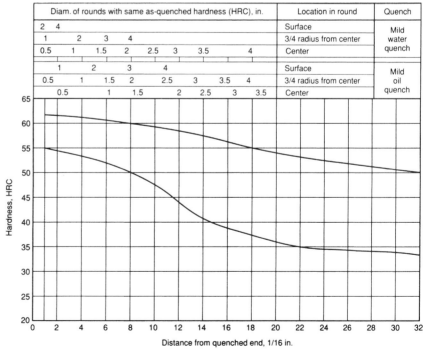

Fig. A7.4 Heat treating temperatures recommended by SAE. Normalize (for forged or rolled specimens only): 870 °C (1600 °F). Austenitize: 845 °C (1550 °F). Source: Ref 1

SAE/AISI 4145H, UNS H41450

Hardness limits for specification purposes

J distance, mm	Hardness, HRC Maximum	Hardness, HRC Minimum
1.5	63	56
3	63	55
5	63	55
7	62	54
9	62	53
11	61	52
13	61	51
15	60	50
20	59	47
25	58	42
30	57	39
35	56	37
40	55	35
45	55	34
50	55	34

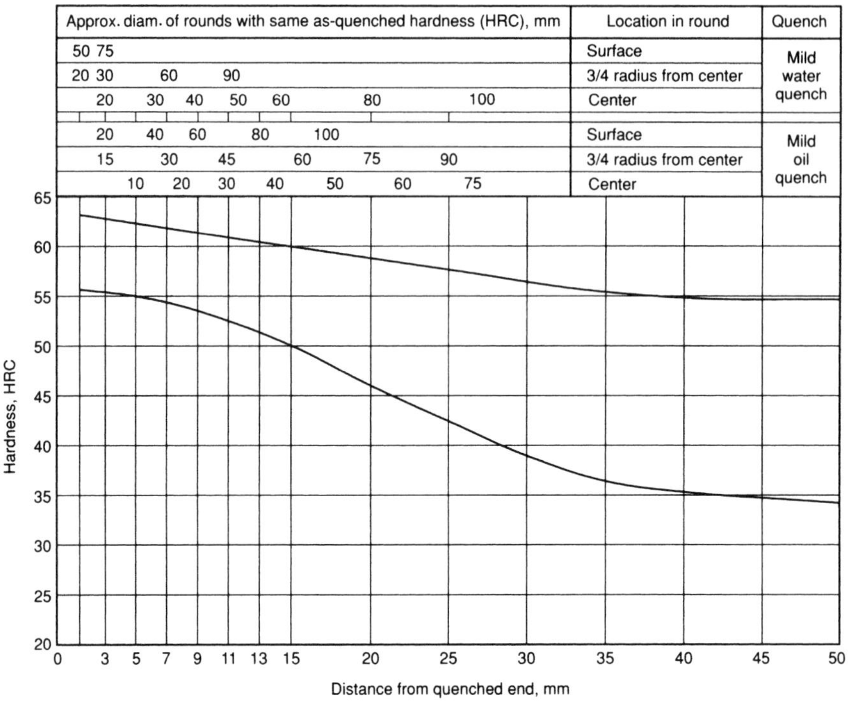

Approx. diam. of rounds with same as-quenched hardness (HRC), mm
Location in round
Quench
50 75
Surface
20 30 60 90
3/4 radius from center
20 30 40 50 60 80 100
Center
Mild water quench
20 40 60 80 100
Surface
15 30 45 60 75 90
3/4 radius from center
10 20 30 40 50 60 75
Center
Mild oil quench
Hardness, HRC
65
60
55
50
45
40
35
30
25
20
0 3 5 7 9 11 13 15 20 25 30 35 40 45 50
Distance from quenched end, mm

Hardness limits for specification purposes

J distance, 1/16 in.	Hardness, HRC Maximum	Hardness, HRC Minimum
1	63	56
2	63	55
3	62	55
4	62	54
5	62	53
6	61	53
7	61	52
8	61	52
9	60	51
10	60	50
11	60	49
12	59	48
13	59	46
14	59	45
15	58	43
16	58	42
18	57	40
20	57	38
22	56	37
24	55	36
26	55	35
28	55	35
30	55	34
32	54	34

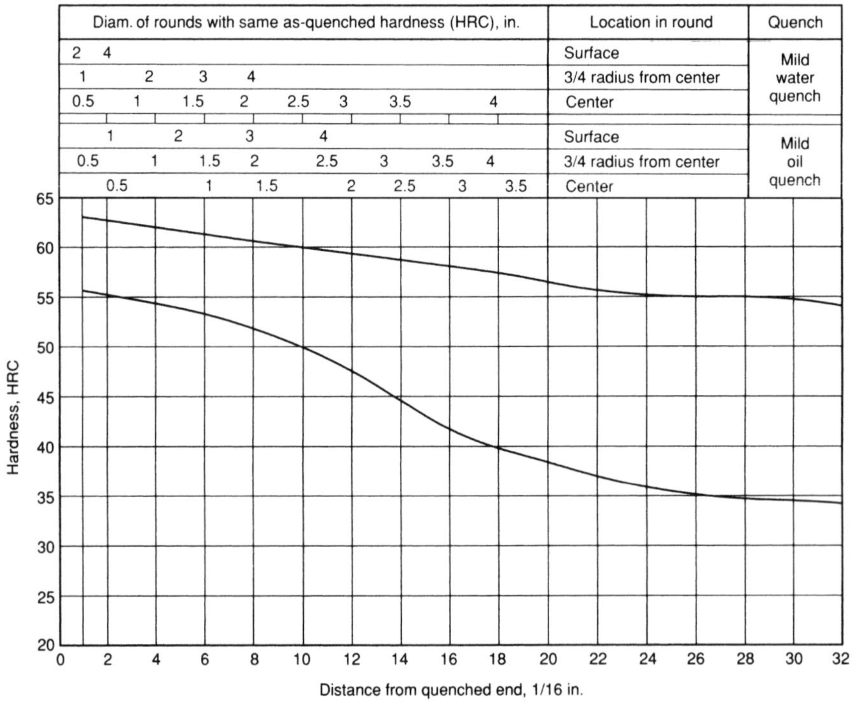

Diam. of rounds with same as-quenched hardness (HRC), in.
Location in round
Quench
2 4
Surface
1 2 3 4
3/4 radius from center
0.5 1 1.5 2 2.5 3 3.5 4
Center
Mild water quench
1 2 3 4
Surface
0.5 1 1.5 2 2.5 3 3.5 4
3/4 radius from center
0.5 1 1.5 2 2.5 3 3.5
Center
Mild oil quench
Hardness, HRC
65
60
55
50
45
40
35
30
25
20
0 2 4 6 8 10 12 14 16 18 20 22 24 26 28 30 32
Distance from quenched end, 1/16 in.

Fig. A7.5 Heat treating temperatures recommended by SAE. Normalize (for forged or rolled specimens only): 870 °C (1600 °F). Austenitize: 845 °C (1550 °F). Source: Ref 1

SAE/AISI 4340H, UNS H43400

Hardness limits for specification purposes

J distance, mm	Hardness, HRC Maximum	Hardness, HRC Minimum
1.5	60	53
3	60	53
5	60	53
7	60	53
9	60	53
11	60	53
13	60	52
15	60	52
20	59	50
25	58	48
30	58	46
35	57	44
40	57	43
45	56	42
50	56	40

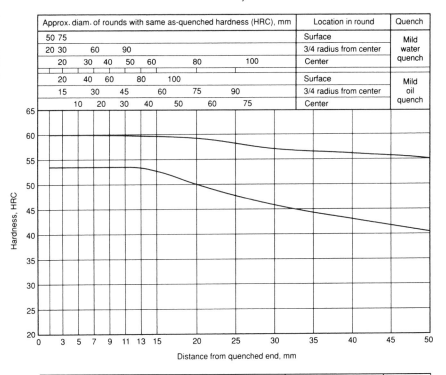

Hardness limits for specification purposes

J distance, 1/16 in.	Hardness, HRC Maximum	Hardness, HRC Minimum
1	60	53
2	60	53
3	60	53
4	60	53
5	60	53
6	60	53
7	60	53
8	60	52
9	60	52
10	60	52
11	59	51
12	59	51
13	59	50
14	58	49
15	58	49
16	58	48
18	58	47
20	57	46
22	57	45
24	57	44
26	57	43
28	56	42
30	56	41
32	56	40

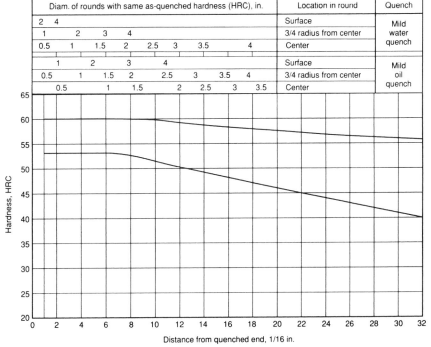

Fig. A7.6 Heat treating temperatures recommended by SAE. Normalize (for forged or rolled specimens only): 870 °C (1600 °F). Austenitize: 845 °C (1550 °F). Source: Ref 1

SAE/AISI 5160H, UNS H51600

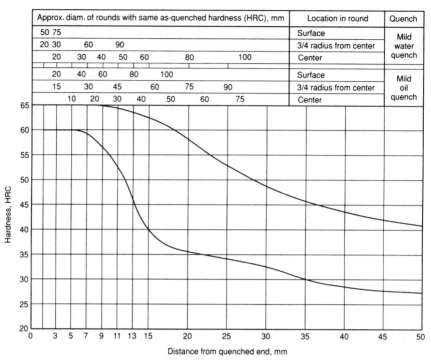

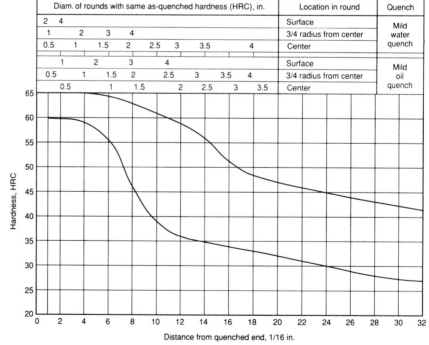

Fig. A7.7 Heat treating temperatures recommended by SAE. Normalize (for forged or rolled specimens only): 870 °C (1600 °F). Austenitize: 845 °C (1550 °F). Source: Ref 1

SAE/AISI 8620H, UNS H86200

Hardness limits for specification purposes

J distance, mm	Hardness, HRC Maximum	Hardness, HRC Minimum
1.5	48	41
3	47	37
5	44	31
7	40	25
9	35	22
11	33	20
13	30	...
15	29	...
20	26	...
25	24	...
30	23	...
35	23	...
40	23	...
45	22	...
50	22	...

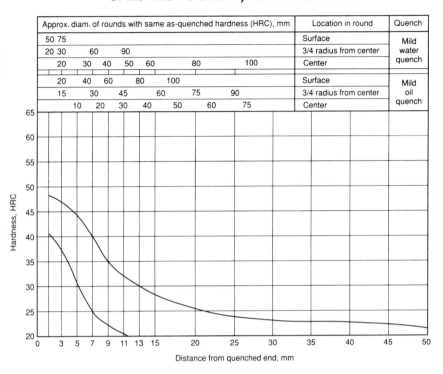

Hardness limits for specification purposes

J distance, 1/16 in.	Hardness, HRC Maximum	Hardness, HRC Minimum
1	48	41
2	47	37
3	44	32
4	41	27
5	37	23
6	34	21
7	32	...
8	30	...
9	29	...
10	28	...
11	27	...
12	26	...
13	25	...
14	25	...
15	24	...
16	24	...
18	23	...
20	23	...
22	23	...
24	23	...
26	23	...
28	22	...
30	22	...
32	22	...

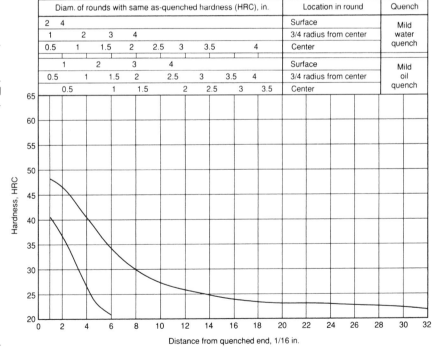

Fig. 7.8 Heat treating temperatures recommended by SAE. Normalize (for forged or rolled specimens only): 925 °C (1700 °F). Austenitize: 925 °C (1700 °F). Source: Ref 1

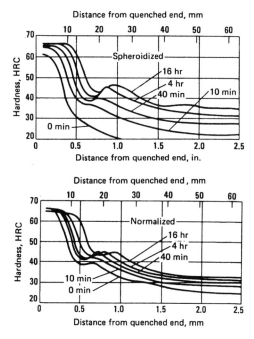

Fig. A7.9 E52100 end-quench hardenability. Austenitized at 845 °C (1550 °F). Insufficient time to permit full carbide solubility in austenite.
Source: Ref 2

REFERENCES

1. *Properties and Selection: Irons and Steels*, Vol 1, *Metals Handbook*, 9th ed., American Society for Metals, 1978
2. Standard Practices and Procedures for Steel, *Heat Treaters Guide*, American Society for Metals, 1982

Index

A

Alloy steels
    austenitizing . . . . 96(T), 97, 191(T)
    classification . . . . 77–78, 77(T)
    defects . . . . 161–163(F)
    hardness after tempering . . . . 140(T), 143(F)
Alternating current
    definition 267
    and electromagnetism 6–8
American Iron and Steel Institute 76
American Society for Metals 1, 2
Analysis. *See* Process qualification and
    analysis
Annealing, definition 86–87, 267
Applications
    automotive . . . . 176–177, 177(F)
    cylindrical bar and rod . . . . 179–180(F), 180
    gears . . . . 169–174, 170(F), 172(F),
                                173(F),174(T), 175
    ordnance 175
    shafts . . . . 166–169, 176(F)
    surface hardening . . . 165, 166–177, 203–209
    through-hardening 165, 177–180
    tools . . . . 175–176, 177(F),
                                    178, 178(F)
    track links . . . . 178–179, 178(F)
ASTM standards 76, 215–216, 217,
                                    233, 235
Austenite
    definition 86, 267
    fast cooling 102–106
    retained 106, 240
    transformation 94–102
Austenitization
    1042 steel, prior structure and
        transformation temperature . . . . 96(F)
    austenitizing temperatures for
        steels . . . . 96(T), 191(T)
    transformation to austenite 94–102
    of various irons and steels 97–98
Austenitizing
    definition 87
    and workpiece nonconformance 237
Axle shafts, bending fatigue . . . . 167(F)

B

Bainite
    definition 103, 105, 267
    lower . . . . 105(F)
Boiling stage, definition 267
Boron steels . . . . 77(T), 78–79
Brinell hardness tester, definition 268
Bus bars . . . . 68–69, 70(F), 71(F), 289

C

Camshaft lobes, hardening of . . . . 177(F)
Capacitors
    definition 268
    maintenance 262
Carbides. *See also* Iron carbides
    definition 268
Carbon
    definition 268
    and hardenability 92–94
    in iron 82–86
    and workpiece nonconformance . . . 237, 245
Carbon restoration, definition 88, 268
Carbon steels
    austenitizing . . . . 96(T), 97, 99–102,
                                        191(T)
    bending fatigue . . . . 167(F)
    classification . . . . 76–77, 77(T)
    continuous bar heating . . . . . . 179(F), 180(F)
    Curie temperature . . . . 10(F)
    decarburization 151–155
    defects . . . . 158(F), 161–163(F)
    definition 268
    hardness after tempering . . . . 140(T), 143(F)
    isothermal transformation diagram . . . . 98(F)
    microstructure . . . . 100(F), 101(F)
    quench cracking 160
Carbonitriding, definition 88, 268
Carburizing, definition 88, 268
Case depth
    effective 193, 231, 232, 270
    magnetic flux concentrators and . . . . . . 48(F)
    minimum 231

Case Depth (continued)
   process qualification 185–186
   specifications 193
   total 193, 231, 276
   and workpiece nonconformance . . . 241, 242
Cast irons
   austenitizing 97
   classification . . . . 79–81, 79(T), 80(F), 80(T)
   definition 268
   mechanical properties . . . . 80(T)
   tensile strength . . . . 79(T)
Caterpillar Tractor Co. 1–2
Cementite. *See also* Iron carbides
   in carbon steel . . . . 82–83, 102(F)
Charpy impact tests, definition 268
Cleaning, of
   workpiece 149–150, 201, 210, 230
Coils, induction. *See* Induction coils
Cold drawn bars, definition 269
Compounds, definition 269
Compression, definition 269
Conduction of heat. *See* Heat conduction
Conductors, definition 269
Continuous cooling diagrams
   complete martensitic
      transformation . . . . 135(F)
   definition 102–103, 269
Convection, definition 269
Convection stage, definition 269
Cooling curves
   4340 steel . . . . 104(F)
   quenchants, effect of . . . . 127(F), 131(F)
Cooling systems
   closed loop recirculation . . . . . . 26(F), 27(F)
   galvanic corrosion 25
   maintenance 263
   quench system design 299–305
   water requirements 24–28
Core hardness, definition 269
Cracking. *See also* Defects; Quench cracking;
   Stress cracking
   detection 232
   in shafts . . . . 242(F)
   and workpiece nonconformance . . . 242–244
Critical frequency
   as function of bar diameter . . . . 13(F)
   and heating efficiency . . . . 14(F)
   for various materials . . . . 11(F)
Critical temperature range, definition 269
Cryogenic treatment, definition 269
Crystals, definition 269
Curie temperature
   for carbon steels . . . . 10(F)
   and current penetration . . . . 12(F)
   definition 269
Current density, definition 269
Cut-off wheel burn
   definition 269
   and workpiece nonconformance . . . 240–241

D

Decarburization
   in castings 154
   and cracking . . . . 243(F)
   definition 151, 270
   depth 152–154
   in eutectoid steel . . . . 153(F)
   and fatigue behavior . . . . 153(F)
   in forgings 154
   in hypereutectoid steel . . . . 152(F)
   limits, for steel forgings . . . . 154(T)
   significance 155
   as source of testing error 230
   stock removal 154
   and workpiece nonconformance . . . 236–237
Defects. *See also* Cracking; Destructive testing;
   Nondestructive testing; Quality control
   as-received 236–248
   in bolt heads and nuts . . . . 158(F)
   in carbon steels . . . . 158(F)
   detection 158–159
   and heat treating practice 160, 164
   oxide inclusions 244
   quench cracks 159, 243–245
   in rolled bars . . . . 157(F)
   seams . . . . 157, 158–159, 158(F)
   stress cracks 159, 243
   types of 155–158
   and workpiece design 159–160
   and workpiece nonconformance . . . 242–244
Definitions, metallurgical 267–277
Destructive testing, advantages and
   limitations . . . . 216(T)
Diagnostic equipment
   capabilities . . . . 35(T)
   control plan flow diagram . . . . 34(F)
Distortion
   definition 116
   in gears . . . . 116(F), 117(F)
   and process selection 199
   shape change 117
   size 116
   and workpiece nonconformance 245
Ductility, definition 270

E

Eddy currents 7
Effective case depth, definition 270
Electrical output, maintenance 262
Electromagnetic, definition 270
Electromagnetism, and alternating
   current 6–8
Energy monitors
   kW-second . . . . 31(F)
   for monitoring process variables 30–33

parameters .... 32(F)
scan speed .... 31(F)
Etch, definition.................... 270
Etching
  laboratory procedures .... 232(F)
  for total case-depth measurement 231
Eutectic, definition................... 270

F

Faraday, Michael 1
Ferrite
  in carbon steel .... 100(F)
  definition 270
Ferrite banding
  definition 270
  and workpiece nonconformance 238
Fixtures. *See also* Workhandling equipment
  design factors 36–37, 200
  index fixtures 207–208
  lift-and-rotate
    mechanism .... 36(F), 202, 206
  maintenance 266
  production examples............ 203–209
  types........................ 201–202
Flashpoint, definition 270
Flaws. *See* Defects
Flux concentrators. *See* Magnetic flux
    concentrators
Flux field, definition.................. 270
Flux intensifiers 46–50
Forging cracks, in bolt heads and
    nuts .... 158(F)
Forgings, decarburization limits....... 154(T)
Frequency
  coil characteristics vs.............. 50–53
  definition 270
  and full power output 35–36
  and gear hardening .... 171(F), 171–173(F),
                         174(T), 205–207
  and hardened depth .... 187(F)
  for induction tempering........... 147(T)
  and pin hardening.............. 203–205
  selection .... 186(T), 188(F), 203–209

G

Galvanic corrosion, in cooling water
    systems........................ 25
Gears
  distortion............... 116(F), 117(F)
  surface hardening .... 169–174, 170(F),
        172(F), 173(F), 174(T), 175, 205–207
Geometry. *See* Part geometry
Grain size
  ASTM..................... 118(F)

austenitic .... 119(F)
definition 117, 271
martensitic 245
measurement.................. 117–119
and quench cracking............... 160
Grains, definition 271
Graphite, definition 271
Grinding crack, definition.............. 271

H

H-steels..................... 77(T), 78
Hammer heads, hardening of .... 177(F)
Hardenability
  carbon and 92–94
  curves......................... 315–324
  definition 271
  overview...................... 90–92
Hardenability bands, definition.......... 271
Hardening. *See also* Scan hardening; Surface
    hardening; Through-hardening
  errors........................ 229(T)
  and residual stresses 114–115
Hardness
  1045 steel .... 93(F), 114(F), 203–205
  4140 steel .... 139(F)
  and carbon content .... 93(F), 185(F)
  gears .... 2(F), 170(F), 205–207
  overview 89
  pins......................... 203–205
  section size and surface hardness..... 94(F)
  of tempered carbon and alloy
    steels .... 140(T)
  and tempering
    temperature .... 141(F), 145(F)
  and tempering time .... 142(F)
Hardness testing
  and case depth 231, 232
  equipment.................... 216–220
  improper test procedures........ 235–236
  Jominy end-quench
    testing......... 91, 91(F), 92(F), 93(F)
  microhardness (Vickers)
    testing 215, 219–220, 230
  overview...................... 89–90
  Rockwell testing............. 90(F), 215,
        216–219, 217(T), 218(T), 219(F), 230,
                              231, 232, 236, 275
  standards.................... 215–220
  workpiece preparation 230
Heat conduction
  and hardened depth .... 17(F)
  theory....................... 16–17
Heat stations 28, 261–262, 271
Heat of steel, definition 271
Heat treating. *See also* Induction heat
    treating history 1–3

Hertz, definition 271
High frequency, definition 272
Holes, and overheating in round bars . . 195(F),
　197(F)
Homogeneous, definition 272
Hot rolled steel, definition 272
Hysteresis
　effect on heating rate . . . . 9(F)
　theory 8–9

I

IGBT. *See* Transistors, isolated gate bipolar
Induce, definition 272
Induction austenitization. *See* Austenitization.
Induction coils
　adapters . . . . 71(F)
　bus bars 68–69, 289
　butterfly coils . . . . 64–65, 64(F)
　channel or hairpin coils . . . 61, 61(F), 62(F),
　　201, 202
　characterization 50–51, 289–290
　classification 43–45
　coil returns for tempering of wheel
　　spindles and hubs . . . . 72(F)
　computer-aided design 293–296
　connection to leads 288–289
　definition 272
　design and fabrication 45–50, 199,
　　283–297
　efficiency . . . . 45–49, 45(T), 286
　electromagnetic end effect . . . . 51(F)
　electromagnetic field . . . . 6(F), 43–45, 44(F)
　flux concentrators 46–50, 291–293
　flux plot . . . . 44(F)
　and frequency 50–53
　internal coils . . . . 62–63, 62(F), 63(F)
　leads and dead spots 51–52, 288
　longitudinal coils . . . . 43–44, 68(F), 201
　machining 288
　maintenance 263
　materials 283–284, 288
　multi-turn solenoid coils . . . . 69(F), 287
　pancake coils . . . . 64, 64(F)
　quenchants, effect on design 290–291
　quick connects . . . . 70, 71(F)
　recommended coupling distances . . . . . 46(T)
　reverse-turn coils . . . . . . 66–67, 66(F), 67(F)
　for scanning or progressive hardening . . . 53,
　　54(F), 201, 202, 203(T), 204(T), 205(T)
　for single-shot heating 52–53, 201–202,
　　203(T), 204(T)
　single-tooth gear coils . . . . 69(F)
　solenoid encircling coils . . 53–60, 54–61(F),
　　201
　special coils 67–68
　split-return coils . . . . 65–66, 65(F)
　for static heating . . . . 52, 53(F), 201–202

　tempering coils 70, 72
　theory 6–8
　transverse flux coils . . . . 44, 45(F)
　two-turn coils . . . . 66(F), 67(F), 287
　types 53–72
　wall thickness 286
　water flow 285–286
　and workpiece nonconformance . . . 246–248
　and workpiece proportions . . . . 52(F)
Induction equipment, basic types 2
Induction hardening. *See* Hardening.
Induction heat treating
　advantages 3–4
　applications 165–181, 203–212
　basics 75–120
　cooling water requirements 24–28
　decarburization 151–155
　defects and flaws 155–164
　distortion 116–117
　grain size 117–119
　heating rates, estimation of . . . . 203(T)
　history 1–3
　induction coils 43–73, 283–297
　inspection 220–233
　line utility requirements 24, 246
　load matching 39–40
　maintenance 261–266
　metallurgical definitions 267–277
　nonconformance of workpieces 235–248
　power supplies 19–24, 28–37, 40–42
　problem solving 235–248
　process classification 86–89
　process qualification and analysis . . 183–201
　quality control 249–260
　quenching 37–39, 121–136, 299–305
　residual stresses 106, 109–115
　rust protection 150, 201, 210
　standards 215–220
　systems . . . . 19–42, 20(F)
　tempering 137–148, 307–313
　theory 5–18
　workhandling equipment selection　201–212
　workpiece cleaning　149–150, 201, 210, 230
Induction heaters. *See also* Power supplies
　definition 272
　operator instructions 212
Induction tempering. *See* Tempering
Inspection. *See also* Destructive testing;
　Hardness testing; Nondestructive testing;
　Quality control of workpieces 229–233
International Organization for
　Standardization 249–250
Inverters. *See also* Power supplies;
　Rectifiers; Thyristors; Transistors
　bridge . . . . 23(F)
　direct current source . . . . 22(F)
　tuning 40–42
　types . . . . 21(F)
Iron carbides 82–86

definition 272
and workpiece nonconformance ... 238–239
Irons. *See also* Cast irons
    definition 272
Isothermal transformation diagrams
    4340 steel .... 104(F)
    carbon steel .... 98(F)
    definition 99, 272
Izod impact test, definition 272

L

Load matching .... 39–40, 40(F)
Low frequency, definition 272

M

Macrostructure, definition 272
Magnaflux, definition 272
Magnetic flux concentrators .... 46–50, 47(F),
    48(F), 48(T), 291–293
Maintenance
    capacitors 262
    cooling water systems .... 263, 263(T)
    high-frequency electrical output 262
    induction coils 263
    mechanical systems or fixturing 266
    periodic 265–266
    power supplies and heat stations ... 261–262
    quench system 264–265
    rust on workpieces 265
Manganese, definition 272
Martensite
    definition 105–105, 273
    tempered .... 106, 108(F)
    untempered .... 107(F)
Martensite transformation range,
    definition 273
Medium frequency, definition 273
Melting, definition 273
Metal Powder Industries Federation 82
Microhardness, definition 273
Microstructure
    and austenitizing 95–102
    definition 273
    transformation products .. 103, 105–106, 239
    and workpiece nonconformance 238,
    241–242
Monitors, energy. *See* Energy monitors
MOS FET. *See* Transistors, metal-silicon-
    diode field-effect

N

National Institute of Standards and
    Technology 230

Nitriding, definition 88–89
Nitrocarburizing, definition 89
Nonconformance, of workpieces 235–248
Nondestructive testing
    advantages and limitations .. 216(T), 221(T)
    crack detection 160, 232
    eddy-current testing ...... 226–227, 228(F),
    228(T), 232
    liquid-penetrant inspection ... 224–225, 232
    magnetic particle inspection .. 222, 224, 232
    method comparison .... 221(T)
    overview 220–222
    process selection .... 223(T)
    radiographic inspection 228–229
    for surface defects and seams 158–159
    ultrasonic inspection .... 225, 226(F),
    227(F), 232
Normalizing, definition 87, 273
Notch toughness, 4140 steel .... 139(F)

O

Oil quench, definition 273
Operator instructions 212
Output transformer, definition 273

P

Part geometry
    analysis .... 194–197, 195(F), 197(F)
    and quench cracking 244
Part submission warrant .... 251–253(F)
Pearlite
    in carbon steel ...... 100(F), 101(F), 102(F)
    definition 103, 273
Phase, definition 273
Phase diagrams
    definition 83, 273
    iron-cementite .... 84(F)
    iron-graphite .... 85(F)
Pins, surface hardening 203–205
Pitting, definition 274
Pliers, through-hardened .... 178(F)
Polymer quench. *See also* Quenchants, polymer
    definition 274
Post-production processing 201, 210
Powdered metals
    compositions .... 83(T)
    gears 175
    part formation 81–82
Power density
    selection 14–16
    for surface hardening .... 190(T)
    for through-heating .... 192(T)
Power supplies. *See also* Induction heaters
    basic diagram .... 20(F)

Power supplies (continued)
   comparative efficiencies .... 22(T)
   for induction tempering .... 147(T)
   maintenance 261–262
   operator instructions................ 212
   output at resonant frequency .... 35(F)
   production rates 203–209
   radio frequency oscillators 3, 23, 41–42
   regulation...................... 28–37
   solid state 3, 4, 20–23, 40–41
   tuning......................... 40–42
   types.......................... 19–24
   typical system losses .... 15(T)
   and workpiece nonconformance ... 246–248
Precipitate, definition 274
Problem solving 235–248
Process qualification and analysis
   examples..................... 203–209
   frequency selection ...... 186–189, 186(T), 187(F), 188(F)
   geometrical effects............. 194–197, 195(F), 197(F)
   hardness and case depth ... 185–186, 185(F)
   material and prior heat treatment... 183–185
   operator instructions................ 212
   post-production processing 201, 210
   power density and heating time ... 189–192, 190(T), 191(T), 192(T)
   process selection................ 198–200
   quenchant selection ...... 197–198, 198(T)
   setup instructions and procedures .. 210–212
   specification review 193
   steps .... 184(F)
   tempering 200, 209–210
   tolerance review 192–193
   workhandling equipment selection 201–202
Pyrometers for temperature control..... 33–35

Q

Quadruple-head, skewed-drive roller
   system (QHD) .... 168(F)
Qualification. *See* Process qualification and analysis
Quality control. *See also* Inspection
   control plan .... 255–257(F)
   failure mode and effects analysis
     (FMEA) 254, 258, 259
   formalized programs............ 249–258
   malfunction check sheet....... 259–260(T)
   model hardening system... 258(T), 259–260
   production part approval process
     (PPAP) 251, 254
Quench cracking. *See also* Defects
   in steels .... 161–163(F)
   and workpiece nonconformance ... 243–245
Quenchants
   brine-based .... 125, 126(F)
   cooling curves 131
   for crack-susceptible steels .... 198(T)
   and induction coil design 290–291
   oil...................... 125, 127, 149
   polymer..... 127–130, 129(F), 130(F), 150, 239, 264–265, 266, 266(F), 274
   quench severities .... 124(F)
   removal................... 149–150, 230
   selection .... 134–135, 197–198, 198(T)
   types........................ 123–130
   water......... 123–124, 124(F), 125(F), 149, 239, 264, 265, 266
Quenching. *See also* Quenchants
   basic arrangements .... 134(F)
   definition 87, 121, 274
   of hot steel rod.................. 122(F)
   methods 132–134
   stages 121–123
   system design 37–39, 299–305
   system maintenance 264–265
   vapor pocket formation.............. 239
   and workpiece nonconformance... 236–237, 239–240

R

Rectifiers, silicon controlled (SCR) ..... 21(F)
Reference depth
   and object thickness............... 12(F)
   and skin effect 9–14
   for various materials .... 10(F)
Residual stresses
   in carbon steel cylinder ..... 113(F), 115(F)
   in cold drawn steel .... 110, 111(F), 112
   and hardening................ 114–115
   and induction heat treating ... 106, 109–115, 114(F)
   and manufacturing processes ... 109(T), 110
   measurement.................. 109–110
   stress relieving 110, 112
   thermal stresses................ 112–113
   volume changes 113–114
Resistance
   definition 274
   theory......................... 5–6
Resistivity, of various metals........... 8(T)
Resulfurized steel, definition 274
Retained austenite, definition 274
Rust protection 150, 201, 210, 265

S

Sauveur, Albert 1
Scale, definition 275
Scan, definition 275
Scan hardening .... 54(F), 201–202, 208–209, 279–281

Index / 331

SCR. *See* Resistors, silicon controlled
Seams. *See also* Defects
   definition 275
Secondary quench, definition 275
Segregation, definition 275
Setup procedures 210–212
Shafts
   cracks .... 242(F), 243(F)
   diameter changes................ 200(F)
   fatigue.................. 167(F), 168(F)
   surface hardening .... 166–169, 175, 176(F)
   ultrasonic inspection .... 227(F)
Single shot, definition 275
Skin effect, and reference depth 9–14
Society of Automotive Engineers 76
Sorby, Henry Clifton 1
Specific heat, relationship to temperature for
   various materials .... 7(F)
Specifications. *See also* Standards
   review of 193
Spheroidical, definition 275
Spheroidizing, in carbon steel .... 101(F)
Spotty hardness, definition 275
Stainless steels
   austenitizing.............. 97–98, 191(T)
   classification .... 77(T), 78(F), 79
   defects....................... 163(F)
   definition 275
Standards. *See also* ASTM standards;
   Specifications................ 215–220
Static heat cycle, definition............. 275
Steels. *See* Alloy steels; Boron steels; Carbon
   steels; H-steels; Powdered metals;
   Stainless steels; Tool steels
Stress, definition 276
Stress cracking. *See also* Defects 159, 243
Stress relieving
   of cold-drawn bars 112
   definition 87–88, 276
   of residual stress............... 110, 112
Stress risers 159–160
Surface hardening
   applications 165, 166–177, 203–209
   required power density.... 189–190, 190(T)

T

Temperature control
   pyrometers for 33–35
   simple system .... 33(F)
Tempered martensite, definition 276
Tempering
   batch tempering 142–144
   continuous furnaces 144
   and crack reduction 164
   curves..................... 311–313
   definition 87, 276
   dimensional changes................ 139

   furnace tempering................. 142
   Hollomon-Jaffe correlation 307–310
   induction tempering 144–147
   overview.................... 137–138
   power source and frequency
     selection .... 147(T)
   processes 142–148
   progressive, operating and production
     data .... 148(T)
   residual heat tempering............. 148
   short-time...................... 307
   specifications 200, 209–210
   stages 138–139
   tempering temperature, effect of .... 138(F),
            140–141, 141(F), 143(F)
   tempering time, effect of .. 141–142, 142(F)
Tension, definition 276
Through-hardening
   applications............... 165, 177–180
   required power density.... 190–192, 192(T)
Tolerances, review of 192–193
Tool steels
   austenitizing................... 97–98
   classification .... 81, 82(T)
   definition 276
Torsion, definition 276
Total case depth, definition............. 276
Track links, through-hardened .... 178(F)
Transformation product, definition 276
Transformation range, definition......... 276
Transformation temperature, definition.... 276
Transformer, definition................. 276
Transistors
   isolated gate bipolar (IGBT) .... 21(F)
   metal-silicon-diode field-effect
     (MOS FET) .... 21(F)
Transition product, definition 276
Transmission shafts, fatigue life....... 168(F)
Tuning, of solid-state inverters 40–42

U

Unified Numbering System 76
Utility line requirements............ 24, 246

V

Vapor phase, definition................. 277
Vapor pocket formation, and workpiece
   nonconformance................. 239
Vickers, definition 277

W

Warrant. *See* Part submission warrant

Water
    in cooling systems 24–28
    as quenchant . . . . . 123–124, 124(F), 125(F)
Work coil, definition 277
Workhandling equipment. *See also* Fixtures
    design factors 36–37, 200
    selection 201–212
Workpiece geometry. *See* Part geometry
Workpiece nonconformance. *See* Nonconformance

AUG 3 0 2007 ✓
OCT 1 8 2008
MAY 2 2 2018